ASVAB STUDY GUIDE 2024-2025

ASVAB Prep Book with 7 Practice Tests, Digital Flashcards with Proven Strategies to Ace your Exam for Air Force, Army, Coast Guard, Marine Corps, National Guard, Navy

Copyright 2023 – PrepGenius - **All rights reserved.**

The content contained within this book may not be reproduced, duplicated, or transmitted without direct written permission from the author or the publisher.

Under no circumstances will any blame or legal responsibility be held against the publisher or author for any damages, reparation, or monetary loss due to the information contained within this book, either directly or indirectly.

Legal Notice:

This book is copyright-protected. It is only for personal use. You cannot amend, distribute, sell, use, quote, or paraphrase any part of this book's content without the author's or publisher's consent.

Disclaimer Notice:

Please note the information contained within this document is for educational and entertainment purposes only. All effort has been executed to present accurate, up-to-date, reliable, and complete information. No warranties of any kind are declared or implied.

Readers acknowledge that the author does not render legal, financial, medical, or professional advice. The content within this book has been derived from various sources. Please consult a licensed professional before attempting any techniques outlined in this book.

By reading this document, the reader agrees that under no circumstances is the author responsible for any direct or indirect losses incurred due to the use of the information contained within this document, including, but not limited to, errors, omissions, or inaccuracies.

PREPGENIUS

Empowering Minds,
One test at a Time

PrepGenius.IO/ASVAB

Create your free account and get access:
- **5 Practice Test**
- **Flashcards**

FREE FLASHCARDS LEARNING SYSTEM

COVERING ALL TEST SUBJECTS

- General Science (GS): Tests knowledge of physical and biological sciences.
- Arithmetic Reasoning (AR): Assesses ability to solve basic arithmetic problems.
- Word Knowledge (WK): Measures the ability to understand the meaning of words through synonyms.
- Paragraph Comprehension (PC): Tests the ability to obtain information from written material.
- Mathematics Knowledge (MK): Measures knowledge of mathematical concepts and applications.
- Electronics Information (EI): Assesses knowledge of electrical equipment and electronics.
- Auto & Shop Information (AS): Tests automotive maintenance and repair knowledge and wood and metal shop practices.
- Mechanical Comprehension (MC): Measures understanding of basic mechanical principles and mechanisms.

over 2000 flashcards devided by test subject

support@prepgenius.io

prepgenius.io

Contents

Chapter 1: Introduction to the ASVAB 1
    What is the ASVAB? 1
    Why is it Important? 1
    Looking at How ASVAB Results Affect Military Training Programs and Employment: 1
    Understanding Line Score Calculation for Each Branch: 2

Chapter 2: ASVAB Test Sections 5
    ASVAB Subtests and What They Measure: 5
    Tips for Each Subtest: 5
    Registering for and Taking the ASVAB: 6
    How to read your ASVAB Scores? 6
    Understanding the Four Major Categories: Your AFQT Results: 7
    Calculating the AFQT Score: 7
    Looking at AFQT Score Requirements for Enlistment: 7
    Do-Over: Retaking the ASVAB: 9
    Army Retesting Policy in the United States: 9
    Policy on retesting in the United States Air Force: 9
    Policy on retesting in the United States Navy: 9
    Policy on retesting in the United States Marine Corps: 10
    Policy on retesting in the United States Coast Guards: 10

Chapter 3: ASVAB Study Strategies 11
    Setting Goals and Creating a Study Plan: 11
    Best Practices for Studying Effectively: 12
    Taking the Exam: Computerized or Paper? 13
    The Advantages and Disadvantages of Computerized Testing: 14
    The Benefits and Drawbacks of the Paper Version: 15
    Dealing with Multiple-Choice Questions: 15
    Intelligent Guessing: 16
    Preparing for the ASVAB through Study and Practice: 17
    Making Your Study Schedule: 18
    Outlining Your ASVAB Study Strategy: 19
    Planning Your Review Three Weeks in Advance: 19

Chapter 4: ASVAB Test Day 21
    What to Expect on Test Day? 21
    Tips for Staying Calm and Focused: 21
    Strategies for Answering Questions Efficiently 21

Chapter 5: General Science (GS) 23
PART I: LIFE SCIENCE: 23
Human Body Systems and Diseases THE SKELETON AND MUSCLES 23
THE RESPIRATORY SYSTEM: 24
BLOOD AND THE CIRCULATORY SYSTEM: 24
The Human Heart: 25
The Capillary System: 26
THE DIGESTIVE AND EXCRETORY SYSTEMS: 26
The Digestive System: 27
THE NERVOUS SYSTEM 27
THE REPRODUCTIVE SYSTEM: 28
HUMAN PATHOGENS: 28
Genetics: 29
Cellular Structures and Functions: 30
Ecology: 31
The Classification of Living Things: 32
Life Science Practice Questions: 33

Part II EARTH AND SPACE SCIENCE: 34
GEOLOGY: 34
STRUCTURE OF EARTH: 34
PLATE TECTONICS: 35
GEOLOGIC TIME SCALE: 35
Cycles in Earth Science: 35
Meteorology: 36
EARTH'S ATMOSPHERE: 36
FRONTS: 37
CLOUDS: 37
Our Solar System: 37
THE SUN: 37
THE PLANETS AND OTHER PHENOMENA: 38
Earth and Space Science Practice Questions: 39
Answers and Explanations EARTH AND SPACE SCIENCE PRACTICE QUESTIONS: 40

Part III PHYSICAL SCIENCE: 40
Measurement: 41
Physics: 42
MOTION 42
FORCES AND ENERGY: 43
NEWTON'S LAWS: 43
ENERGY: 44
SOUND WAVES: 44

Contents | vii

    THE ELECTROMAGNETIC SPECTRUM: 45

    OPTICS: 45

    HEAT: 46

    MAGNETISM: 46

    Chemistry: ELEMENTS AND THE PERIODIC TABLE: 46

    COMPOUNDS: 48

    ACIDS AND BASES: 48

    PHYSICAL CHANGE: 49

    CHEMICAL CHANGE: 49

    Physical Science Practice Questions: 49

    Answers and Explanations PHYSICAL SCIENCE PRACTICE QUESTIONS: 50

Part IV GENERAL SCIENCE PRACTICE QUESTIONS: **51**

    GENERAL SCIENCE PRACTICE SET 1: 51

    Answers and Explanations: 53

    GENERAL SCIENCE PRACTICE SET 2: 54

    REVIEW AND REFLECT: 56

Chapter 6: Arithmetic Reasoning (AR) **57**

    Tackling the Real World of Word Problems: 57

    Reading the Entire problem: 57

    Figuring Out What the Question is Asking: 57

    Digging for the Facts: 58

    Setting Up the Problem and Working Your Way to the Answer: 58

    Drawing a diagram 59

    Reviewing your answer: 59

    The Guessing Game: Putting Reason in Your Guessing Strategy 59

    Using the Process of Elimination: 60

    Solving What You Can and Guessing the Rest: 60

    Making Use of the Answer Choices: 60

    Arithmetic Reasoning (Math Word Problems) Practice Questions 61

    Easy Arithmetic Reasoning questions: 61

    Answer Key with Explanations 63

Chapter 7: Word Knowledge (WK) **65**

    Understanding How Important It Is to Know Words: 65

    Trying Out the Word Knowledge Question Format: 66

    How to Build Words from Scratch: Tips to Help You Figure Out What Words Mean: 66

    From the Start to the End, Knowing Prefixes and Endings are Important: 66

    Getting to the Bottom of a Problem: 68

    Word Families: Finding Similar Things: 69

    Taking Words Apart: 70

    How to Use Synonyms and Antonyms: "Yin and Yang" 70

ASVAB Word Knowledge Strategy: What to Do When You Don't Know the Answer:....................................71
Making Your Situation:....................................71
What's it Like? Letting the Words Tell You What to do:....................................71
Getting Clues: A Comparison and Difference:....................................72
One of These is Not the Same as the Other: Getting Rid of the Wrong Answers:....................................72
Care to Fill in? Change the Word to One of the Choices:....................................72
Parts of Speech:....................................73
If Nothing Else Works, Break Your Word:....................................74
Improve Your Vocabulary and Yourself:....................................74
You Can Learn More Words by Reading:....................................74
Making a List and Double-Checking it:....................................75
Crosswords are a Fun Way to Learn Words:....................................75
Questions to Test Your Word Knowledge:....................................75
Simple Questions to Test Your Word Knowledge:....................................76
Medium Word Questions About Knowledge and Practice:....................................76
Questions to Test Your Knowledge of Hard Words:....................................77
Answers and Breakdowns:....................................78

Chapter 8: Paragraph Comprehension (PC)....................................81
Conquering Paragraph Comprehension Questions: Global Questions....................................81
INFERENCE QUESTIONS:....................................82
Vocabulary-in-Context Questions:....................................82
Paragraph Comprehension Practice Set 1:....................................82
Paragraph Comprehension Practice Set 2:....................................84
Answers and Explanations
PARAGRAPH COMPREHENSION PRACTICE SET 1:....................................86
PARAGRAPH COMPREHENSION PRACTICE SET 2:....................................87

Chapter 9: Mathematics Knowledge (MK)....................................91
Operations: What You Do to Numbers:....................................92
First Things First: Following the Order of Operations:....................................92
Completing a Number Sequence:....................................93
Finding the Pattern:....................................93
Dealing with More than One Operation in a Sequence:....................................94
Averaging Mean, Median, and Mode in a Range:....................................94
Working on Both Sides of the Line: Fractions....................................94
Common Denominators: Preparing to Add and Subtract Fractions:....................................95
Multiplying and Reducing Fractions:....................................96
Dividing Fractions:....................................96
Converting Improper Fractions to Mixed Numbers and Back Again:....................................97
Expressing a Fraction in Other Forms: Decimals and Percents....................................97
Using Ratios to Show Comparisons:....................................98
Writing in Scientific Notation:....................................99

Tips for Taking Math Tests to Help You on Your Mathematical Journey: 99
Knowing What the Question is Asking: 100
Getting Clear on What You're Trying to Solve: 100
Trying to Figure out What You Can and Guessing the Rest: 100
Using the Process of Elimination: 101
Questions for Practicing Math Skills and Operations: 101
ASVAB Mathematics Knowledge Practice Questions 101
ASVAB Mathematics Knowledge Practice Questions - Set 2 102
ASVAB Mathematics Knowledge Practice Questions - Set 3 103
ASVAB Mathematics Knowledge Practice Questions - Set 4 103
ASVAB Mathematics Knowledge Answer Key 104
ASVAB Mathematics Knowledge Answer Key - Set 2 104
ASVAB Mathematics Knowledge Answer Key - Set 3 105
ASVAB Mathematics Knowledge Answer Key - Set 4 105

Chapter 10 : Electronics Information (EI) 107

Electron Flow Theory: 107
Current: 108
Voltages: 108
Resistance: 108
EXCEPTIONS AND ADDITIONAL NOTES: 109
Circuits: 109
OHM'S LAW: 109
SERIES CIRCUITS: 110
PARALLEL CIRCUITS: 111
SERIES-PARALLEL CIRCUITS: 112
ELECTRICAL POWER: 112
STANDARD ELECTRICAL UNITS AND THE METRIC SYSTEM: 112
Structure of Electrical and Electronic Systems: 113
AC VS. DC: 113
GROUND: 113
IMPORTANT ELECTRIC AND ELECTRONIC COMPONENTS: 113
Resistors: 113
Fuses and Circuit Breakers: 114
Capacitors: 114
SEMICONDUCTORS: 115
Diodes: 115
Transistor: 115
Electricity and Magnetism: 116
INDUCTORS: 116
TRANSFORMERS: 117
BASIC ELECTRICAL MOTORS AND GENERATORS: 117

Electronics Information Practice Set 1: 117

ELECTRONICS INFORMATION PRACTICE SET 2: 119

Answers and Explanations
ELECTRONICS INFORMATION PRACTICE SET 1 120

ELECTRONICS INFORMATION PRACTICE SET 2 122

Chapter 11 : Automotive and Shop Information (AS) 125

Engine Systems: 125

Components: 125

COOLING SYSTEM: 126

Coolant: 126

Components 126

Maintenance 127

LUBRICATION SYSTEM: 127

Engine Oil: 127

Components: 128

OPERATION: 128

Combustion Systems: FUEL SYSTEM: 128

Fuel Injection System Designs: 129

Maintenance: 129

Components: 129

Primary Ignition Operation: 130

Secondary Ignition Operation: 130

Current Ignition System Design: 130

EXHAUST SYSTEMS: 131

Components: 131

Operation: 131

Electrical and Control Systems: ELECTRICAL SYSTEM: 132

Subsystems: 132

SUSPENSION AND STEERING SYSTEM: 133

BRAKE SYSTEMS: 134

Auto Information Practice Set 1 134

Auto Information Practice Set 2: 135

Answers and Explanations
AUTO INFORMATION PRACTICE SET 1: 135

AUTO INFORMATION PRACTICE SET 2 136

Chapter 12 : Mechanical Comprehension 139

A Review of the Physics of Mechanical Devices: 139

MASS AND FORCE: 139

NEWTON'S LAWS OF MOTION: 139

Common Types of Forces:
GRAVITY AND WEIGHT: 140

FRICTION: 140
TENSION: 141
HYDRAULIC PRESSURE: 141
TORQUE: 142
Energy, Work, and Power: 142
WORK: 143
POWER: 143
Simple Machines: 143
INCLINED PLANE: 144
WEDGE: 144
LEVERS: 144
PULLEYS: 145
BLOCK AND TACKLE: 145
WHEEL AND AXLE: 145
GEARS AND GEAR RATIOS: 145
Mechanical Comprehension Practice Set 1: 146
Mechanical Comprehension Practice Set 2 147
Answers and Explanations
MECHANICAL COMPREHENSION PRACTICE SET 1 149
MECHANICAL COMPREHENSION PRACTICE SET 2 150

Chapter 13 : Assembling Objects (AO) 153
Understanding the Concept of Object Assembling: 153
Two Types of Questions for the Price of One: 153
Putting Tab A into Slot B: Connectors: 153
Solving the Jigsaw Puzzle: Shapes: 154
Tips for the Assembling Objects Subtest: 155
Comparing one piece or point at a time: 155
Practicing Spatial Skills: 156
Assembling Objects Practice Questions: 156
Answers and Explanations: 157

CHAPTER 14 : Practice Exam 1: 159
General Science: 159
Arithmetic Reasoning: 160
Word Knowledge: 163
Paragraph Comprehension: 165
Mathematics Knowledge: 167
Electronics Information: 169
Auto & Shop Information: 170
Mechanical Comprehension: 171
Assembling Object: 174

Practice Exam 1 176

Answers and Explanations 176

    General Science Answers... 176

    Arithmetic Reasoning Answers 177

    Word Knowledge Answers: 179

    Paragraph Comprehension Answers: 181

    Mathematics Knowledge Answers: 182

    Electronics Information Answers 184

    Auto & Shop Information Answers: 185

    Mechanical Comprehension Answers: 186

    Assembling Objects Answers: 188

CHAPTER 15: Practice Exam 2 191

    General Science: 191

    Arithmetic Reasoning: 192

    Word Knowledge: 194

    Paragraph Comprehension: 196

    Mathematics Knowledge: 199

    Electronics Information: 200

    Auto & Shop Information: 202

    Mechanical Comprehension: 203

    Assembling Objects: 206

    Practice Exam 2: Answers and Explanations 209

    General Science Answers: 209

    Arithmetic Reasoning Answers: 211

    Word Knowledge Answers: 212

    Paragraph Comprehension Answers: 214

    Mathematics Knowledge: 215

    Electronics Information Answers: 217

    Auto & Shop Information Answers: 218

    Mechanical Comprehension Answers: 219

    Assembling Objects Answers: 221

CHAPTER 1

Introduction to the ASVAB

What is the ASVAB?

The ASVAB (Armed Services Vocational Aptitude Battery) is the world's most widely used multiple-aptitude test battery. It measures a test taker's suitability to enlist in the United States Armed Forces and assesses their abilities to be trained in specific civilian or military jobs.

When you take the ASVAB officially, you will be given either a paper-and-pencil version of the test or a computer version (also referred to as a Computer Adaptive Test ASVAB or CAT-ASVAB). About 70 percent of prospective recruits take the CAT-ASVAB, which will likely be the version you'll encounter.

In this book, you'll learn how the ASVAB is structured and what it tests. You'll also learn strategies and methods that will help you improve your score significantly in each section.

Why is it Important?

The ASVAB is an essential test for individuals interested in enlisting in the military because it helps determine their eligibility for various military jobs and careers. The test measures an individual's aptitude and abilities in multiple areas, which can be used to match them with appropriate military occupational specialties (MOS).

Additionally, the ASVAB plays a crucial role in determining an individual's eligibility for enlistment in the military. The AFQT score, derived from the ASVAB scores, determines if an individual meets the minimum qualifications for enlistment. The minimum score required for enlistment varies depending on the branch of the military and the specific job or career field.

The ASVAB is a vital test providing valuable information for individuals and the military. It helps individuals determine their strengths and weaknesses, and it helps the military match individuals with appropriate job opportunities. It helps ensure that only qualified individuals can enlist in the military.

Looking at How ASVAB Results Affect Military Training Programs and Employment:

Each military branch has a unique scoring system. These scores and other elements, including job availability, eligibility for security clearances, and medical requirements, are used by recruiters and military career counselors to connect prospective recruits with military jobs.

Your service branch chooses your military profession or enlistment program during the first enlisting process based on minimum line scores, which are different combinations of results from specific subtests (for more information, see the next section).

As long as the job you desire is open and you meet other requirements, you can acquire it if you receive high enough marks in the appropriate categories. Except for those joining the infantry or trying out for Special Forces, the Army is the only service that considers test results and provides a job guarantee for all recruits.

In other words, almost every Army recruit knows their position before signing the enlistment contract. The other active duty services combine promised employment opportunities or assured aptitude and professional fields:

>> Air Force: A job guarantee is offered to about 40% of active service Air Force recruits. Most recruits enlist in one of four aptitude areas guaranteed, and during basic training, these recruits are given jobs related to their aptitude area.

>> **Coast Guard:** In its active duty enlistment contracts, it only sometimes provides an employment guarantee. Instead, fresh recruits to the Coast Guard enlist as undesignated seamen and serve for about a year doing general labor ("Paint that ship!") before deciding to seek specialized job training.

>> **Marine Corps:** The vast majority of Marine Corps active duty recruits are assured employment in one of several occupations, including infantry, avionics, logistics, vehicle maintenance, aircraft maintenance, munitions, and so forth. Military Occupational Specialties (MOSs) are subdivided into specific sub-jobs within each discipline. It's common for Marine recruits to discover their true MOSs until halfway through basic training.

>> **Navy:** Most Navy recruits enlist with a guaranteed job; however, several hundred persons enroll in a career field each year with a job guarantee and strike (application) for the particular position within a year of boot camp.

Understanding Line Score Calculation for Each Branch:

The A-line score combines standard ASVAB scores to determine which positions or training programs you are eligible for. Your results on each ASVAB subtest (with Word Knowledge and Paragraph Comprehension aggregated as a Verbal Expression score) are your standard scores.

- » Mechanical Comprehension (MC)
- » Electronics Information (EI)
- » Assembling Objects (AO)
- » General Science (GS)
- » Arithmetic Reasoning (AR)
- » Auto & Shop Information (AS)
- » Mathematics Knowledge (MK)
- » Mechanical Comprehension (MC)
- » Word Knowledge (WK), Paragraph Comprehension (PC), and Verbal Expression (VE)

The way that the various armed forces calculate their line scores varies. Some computations include scores for the defunct ASVAB subtests of Coding Speed (CS) and Numerical Operations (NO), averaged across thousands of test takers. The methods used to calculate each branch's line scores are described in the following sections.

Line Scores and the Army

To calculate line scores for job qualification, the Army adds up the standard scores from the ASVAB in ten different categories. Line scores and their component ASVAB subtests are shown in the Table below.

Line Score	Standard Scores Used	Formula Used
Clerical (CL)	Verbal Expression (VE), Arithmetic Reasoning (AR), and Mathematics Knowledge (MK)	VE + AR + MK
Combat (CO)	Arithmetic Reasoning (AR), Coding Speed (CS), Auto & Shop Information (AS), and Mechanical Comprehension (MC)	AR + CS + AS + MC
Electronics (EL)	General Science (GS), Arithmetic Reasoning (AR), Mathematics Knowledge (MK), and Electronics Information (EI)	GS + AR + MK + EI
Field Artillery (FA)	Arithmetic Reasoning (AR), Coding Speed (CS), Mathematics Knowledge (MK), and Mechanical Comprehension (MC)	AR + CS + MK + MC
General Maintenance (GM)	General Science (GS), Auto and Shop Information (AS), Mathematics Knowledge (MK), and Electronics Information (EI)	GS + AS + MK + EI
General Technical (GT)	Verbal Expression (VE) and Arithmetic Reasoning (AR)	VE+AR

Mechanical Maintenance (MM)	Numerical Operations (NO), Auto and Shop Information (AS), Mechanical Comprehension (MC), and Electronics Information (EI)	NO+AS+MC+EI
Operators and Food (OF)	Verbal Expression (VE), Numerical Operations (NO), Auto and Shop Information (AS), and Mechanical Comprehension (MC)	VE + NO + AS + MC
Surveillance and Communications (SC)	Verbal Expression (VE), Arithmetic Reasoning (AR), Auto and Shop Information (AS), and Mechanical Comprehension (MC)	VE+AR+AS+MC
Skilled Technical (ST)	General Science (GS), Verbal Expression (VE), Mathematics Knowledge (MK) and Mechanical Comprehension (MC)	GS+VE+MK+MC

Line Scores and the Navy and Coast Guard

The ASVAB's subtest scores and the Verbal Expression (VE) score, which is the sum of Word Knowledge (WK) and Paragraph Comprehension (PC), are used directly by the Navy and Coast Guard.

Recruiters, career counselors, and enlistees can use an enlistee's line score to get a general idea of what professions they could be a good fit for, even though the Navy and Coast Guard don't utilize them for formal employment decisions. For example, to become an Air Traffic Control Specialist, you must get a combined VE, AR, MK, and MC score of 220 or above on the Armed Services Vocational Aptitude Battery.

The ASVAB line scores for the Navy and Coast Guard are listed in Table.

Line Score	Standard Scores Used	Formula Used
Engineman (ENG)	Auto & Shop Information (AS) and Mathematics Knowledge (MK)	AS + MK
Administrative (ADM)	Mathematics Knowledge (MK) and Verbal Expression (VE)	MK + VE
General Technical (GT)	Arithmetic Reasoning (AR) and Verbal Expression (VE)	AR + VE
Mechanical Maintenance (MEC)	Arithmetic Reasoning (AR), Auto and Shop Information (AS), and Mechanical Comprehension (MC)	AR + AS + MC
Health (HM)	General Science (GS), Mathematics Knowledge (MK), and Verbal Expression (VE)	GS + MK + VE
Mechanical Maintenance 2 (MEC2)	Assembling Objects (AO), Arithmetic Reasoning (AR), and Mechanical Comprehension (MC)	AO + AR + MC
Electronics (EL)	Arithmetic Reasoning (AR), Electronics Information (EI), General Science (GS) and Mathematics Knowledge (MK)	AR + EI + GS + MK
Nuclear Field (NUC)	Arithmetic Reasoning (AR), Mechanical Comprehension (MC), Mathematics Knowledge (MK) and Verbal Expression (VE)	AR + MC + MK + VE
Engineering and Electronics (BEE)	Arithmetic Reasoning (AR), General Science (GS), and two times Mathematics Knowledge (MK)	AR + GS + 2MK

Line Scores and the Marine Corps

The Marine Corps computes its three-line scores for job qualification by adding scores from various ASVAB subtests, as the Table shows.

Line Score	Standard Scores Used	Formula Used
Mechanical Maintenance (MM)	General Science (GS), Auto & Shop Information (AS), Mathematics Knowledge (MK) and Mechanical Comprehension (MC)	GS + AS + MK + MC
General Technical (GT)	Verbal Expression (VE) and Arithmetic Reasoning (AR)	VE + AR
Electronics (EL)	General Science (GS), Arithmetic Reasoning (AR), Mathematics Knowledge (MK) and Electronics Information (EI)	GS + AR + MK + E

Line Scores and the Air Force

The U.S. Air Force calculates scaled scores in four aptitude areas known as MAGE (mechanical, administrative, general, and electronics) using standard results from the ASVAB subtests.

The Air Force MAGE scores are calculated as percentiles that compare you to thousands of other test-takers and range from 0 to 99. In other words, if you had a percentile score of 51, you outperformed 50% of the test takers who were utilized to determine the norm.

The four areas, the subtests that were used, and the formula for determining the score for each size are listed in Table. The test scorer changes the score for a specific region to a percentile after calculating the score for that area.

Line Score	Standard Scores Used	Formula Used
Mechanical	General Science (GS), Mechanical Comprehension (MC), and two times Auto & Shop Information (AS)	GS + MC + 2AS
Administrative	Numerical Operations (NO), Coding Speed (CS), and Verbal Expression (VE)	NO + CS + VE
General	Arithmetic Reasoning (AR) and Verbal Expression (VE)	AR + VE
Electronics	General Science (GS), Arithmetic Reasoning (AR), Mathematics Knowledge (MK), and Electronics Information (EI)	GS + AR + MK + EI

Chapter 2
ASVAB Test Sections

ASVAB Subtests and What They Measure:

Here are the nine subtests that make up the ASVAB and what each of them measures:

1. **General Science (GS)** - This subtest measures your knowledge of basic physical and biological sciences, including earth science, chemistry, and biology.
2. **Arithmetic Reasoning (AR)** - This subtest measures your ability to solve mathematical problems, including basic arithmetic, algebra, and geometry.
3. **Word Knowledge (WK)** - This subtest measures your ability to understand the meanings of words and use them correctly in context.
4. **Paragraph Comprehension (PC)** - This subtest measures your ability to understand written passages and answer questions based on the information provided.
5. **Mathematics Knowledge (MK)** - This subtest measures your knowledge of mathematical concepts and their applications, including algebra, geometry, and trigonometry.
6. **Electronics Information (EI)** - This subtest measures your knowledge of basic electronic principles, including circuits, currents, and resistance.
7. **Auto and Shop Information (AS)** - This subtest measures your knowledge of automotive systems, repair, and general shop practices and tools.
8. **Mechanical Comprehension (MC)** - This subtest measures your understanding of basic mechanical principles, including simple machines, levers, pulleys, and gears.
9. **Assembling Objects (AO)** - This subtest measures your ability to visualize how objects can be made and rotated to form new shapes.

Each subtest is scored separately, and the scores are used to determine your overall ASVAB score. Your overall score determines your eligibility for enlistment in the military and which military occupations you may qualify for.

Tips for Each Subtest:

Here are some suggestions for each subtest of the ASVAB:

1. **General Science (GS):** Review basic biology, chemistry, and physics concepts. Focus on topics such as the scientific method, the structure of atoms and molecules, and the properties of different types of energy.
2. **Arithmetic Reasoning (AR):** Practice essential math skills, including addition, subtraction, multiplication, and division—also, practice solving word problems and algebraic equations.
3. **Word Knowledge (WK):** Reading books, newspapers, and articles improves vocabulary. Learn new words and use them in conversation to reinforce your understanding.
4. **Paragraph Comprehension (PC):** Practice reading and understanding passages quickly and accurately. Look for main ideas and supporting details, and pay attention to the tone and purpose of the passage.
5. **Mathematics Knowledge (MK):** Review algebra, geometry, and trigonometry concepts. Practice solving equations, understanding formulas, and working with graphs and charts.
6. **Electronics Information (EI):** Study basic electronics concepts, such as circuits, voltage, and resistance. Review electronic components, such as diodes, capacitors, and transistors.
7. **Auto and Shop Information (AS):** Review basic automotive and shop concepts, such as engine components, electrical systems, and tool usage. Practice identifying different types of tools and understanding their uses.

8. **Mechanical Comprehension (MC):** Practice understanding basic mechanical principles, such as leverage, pulleys, and gears. Study diagrams and schematics to better understand how different mechanical systems work.
9. **Assembling Objects (AO):** Practice visualizing how objects can be assembled and rotated to form new shapes. Look for patterns and relationships between different things to help you solve problems.

Remember to pace yourself during the test and read each question carefully before answering. Use the process of elimination to narrow down your choices if you're unsure of an answer. Finally, remember to get plenty of rest and eat healthy before taking the test to help you perform your best.

Registering for and Taking the ASVAB:

Your first step toward signing up to take the ASVAB is to contact the local recruiter for the service branch you want to join unless you are taking the ASVAB at your high school as part of the Department of Defense Career Exploration Program.

Your enlistment application will be completed with assistance from the recruiter; you will be required to submit the necessary paperwork. The recruiter will schedule either a proctored ASVAB test or the PiCAT, an un-proctored version of the ASVAB, once you have satisfied the prerequisites for enlistment.

The Pending Internet Computerized Adaptive Test (PiCAT) is a full-length, unsupervised ASVAB you can take whenever convenient. You'll be registered and given an access code by your local recruiter. After you've completed the test, your recruiter can also give you your score. Then, if you decide to enlist, you will take a proctored verification test to confirm your PiCAT score, which will take only 25 to 30 minutes and is much shorter than the ASVAB.

The Tailored Adaptive Personality Assessment System (TAPAS), a proposed new military admission exam, has been piloted by the Armed Forces. Based on the research done thus far, it is hypothesized that this computer-adaptive personality test would reveal information about the motivation of recruits and other noncognitive traits that will show whether they will be successful following enrollment.

As of this writing, the 120-question TAPAS is still available to applicants taking the ASVAB to continue assessing the test's effectiveness. The military wants to find candidates who will do well despite falling short of the required AFQT scores by using the additional data made available by the TAPAS.

Results from the ASVAB are good for two years. You have one calendar month after your first ASVAB test to retake it. You may take a second retest following the initial one when another calendar month has elapsed. If you've had two retests, you must wait at least six months before retaking the test.

How to read your ASVAB Scores?

Your guidance counselor or recruiter will thoroughly explain your official ASVAB scores in various formats. Though some scores might be more important than others, all scores are essential. Here is a summary of what you will discover.

You will be provided with a single numerical score for the AFQT.

Standard Scores: You will be given a Standard Score for each subtest. Your raw results are compared to the raw scores of a representative national sample to determine these scores. The Department of Defense reports that only around 16 percent of people obtain a Standard Score of 59 or above, whereas roughly half of the population achieves a score of 50 or above.

Service Composite Scores: These score combinations, also known as Line Scores, assess a test-takers vocational aptitude and evaluate whether they possess it to the level necessary to be trained for various occupational assignments across all military branches.

For instance, a Navy Engineering Aid composite score (abbreviated EA) results from adding the effects for Arithmetic Reasoning, General Science, and Mathematics Knowledge multiplied by two. An enlistee must obtain a minimum composite score for the desired occupation to be eligible for job training. For another illustration, you must get a specific score that combines the results of the General Science, Arithmetic Reasoning, Mathematics Knowledge, and Electronics Information tests to be eligible for Army electronics training and employment. Contact your local recruiter or go to the webpage for your chosen service branch for further information.

Career Exploration Scores: Three Career Exploration Scores in the composite areas of Verbal Skills, Math Skills, and Science and Technical Skills are given to students who take the ASVAB in their high schools as part of the Career Exploration Program. These scores are reported as standard and percentiles based on gender, grade level, and school.

Understanding the Four Major Categories: Your AFQT Results:

There is no overall score for the ASVAB. When you hear someone remark, "I got an 80 on my ASVAB," they are not referring to their total ASVAB score but their percentile on the Armed Forces Qualification Test (AFQT) score. Only four of the subtests are used to calculate the AFQT score, which determines whether you even qualify to enlist in the military:

- Word Knowledge (WK),
- Paragraph Comprehension (PC),
- Arithmetic reasoning (AR),
- Mathematics Knowledge (MK) is an example of cognitive skills.

The military demands a specific combination of line scores, which may include the scores you receive on the AFQT for every job, from food service roles to specialty occupations in the medical area. The AFQT subtests are only used to determine which occupations you are qualified for.

Determine which areas to concentrate on based on your career objectives. You only need to spend a little time preparing for the Mechanical Comprehension subtest if you're not interested in a profession requiring a high score. Schedule your study time effectively as you prepare for the ASVAB.

Avoid reading the chapter in this book titled "Assembling Objects" if you don't need to worry about the Assembling Objects subtest. Give word knowledge or arithmetic reasoning sometimes. Nevertheless, it's better to read this book and take the practice tests, focusing on all areas of the ASVAB if you don't have a chosen job or are unsure of your possibilities. By performing well on each subtest, you'll increase your pool of potential employers and your appeal as a candidate.

Calculating the AFQT Score:

The military brass (or at least its computers) determines your AFQT score through a very particular process:

1. Add the value of your Word Knowledge score to your Paragraph Comprehension score.
2. Convert the result of Step 1 to a scaled score ranging from 20 to 62. This score is known as your *Verbal Expression* or VE score.
3. To get your raw AFQT score, double your VE score and add your Arithmetic Reasoning (AR) score and your Mathematics Knowledge (MK) score. The basic equation looks like this: Raw AFQT Score = 2VE + AR + MK
4. Convert your raw score to a percentile, which compares your results to thousands of other ASVAB test-takers. For example, a score of 50 means that you scored as well as or better than 50 percent of the individuals the military is comparing you to.

Looking at AFQT Score Requirements for Enlistment:

AFQT scores are grouped into six main categories based on the percentile score ranges in Table. Categories III and IV are divided into subgroups because the services sometimes use this chart for internal tracking purposes, enlistment limits, and incentives. For example, based on your scores, the military decides how trainable you may be to perform jobs in the service.

Category	Percentile Score	Trainability
I	93–99	Outstanding
II	65–92	Excellent
III A	50–64	Above Average

III B	31–49	Average
IV	10–30	Below average
V	1–9	Not Trainable

The U.S. Congress has mandated that the military cannot accept more than 20% of recruits from Category IV or Category V. To be eligible for enlistment, those without high school diplomas must achieve a score of at least the 31st percentile (i.e., fall into Category III B or higher), and even then, they must possess an alternative credential, such as a General Educational Development (GED) certificate or another high school equivalency certificate.

Because at least 60% of recruits must perform better than average on the AFQT, your chances of being able to enlist are reduced if your score falls in Category III B or anywhere in Category IV (mainly if other Category IV recruits enlist before you do).

The military has varied AFQT score requirements depending on whether you have a high school diploma or a passing score on your state's recognized high school equivalency test (like the GED).

The minimal scores needed for each branch can—and usually do—vary due to the military's shifting requirements. As an illustration, the Army accepted candidates with GEDs and AFQT scores of 31 during the height of Operation Iraqi Freedom.

Branch of Service	Minimum AFQT Score with High School Diploma	Minimum AFQT Score with High School Equivalency Test Certificate	Special Circumstances
Air Force	31	65	The Air Force allows less than 1 percent of its enlistees each year to have a high school equivalency test certificate instead of a high school diploma. Suppose you have a high school equivalency certificate. In that case, you must have at least fifteen hours of college credits to gain the same eligibility as a high school graduate and wait for an applicant slot to become available.
Army	31	60	The Army sometimes approves waivers for applicants with high school equivalency test certificates and AFQT scores below 31.
Coast Guard	40	varies	If you have a high school equivalency certificate, the minimum AFQT score doesn't apply. If your ASVAB line scores qualify you for a specific job and you're willing to enlist in that job, your recruiter may be able to put in a waiver. Very few people (about 5 percent) can enroll with a high school equivalency certificate each year.
Marine Corps	31	50	Sometimes, the Marines issue waivers for people with scores below the minimum thresholds, particularly when the Corps struggles to fulfill enlistment goals.
Navy	31	50	If you enlist with a high school equivalency certificate, you must have 15 college credits.

To enter, you must obtain the minimum AFQT score necessary by each branch; however, the higher your score, the better. For instance, the military officers who make those decisions are more likely to take a chance on you if they think you're a reasonably intelligent cookie than they would be if you barely met the minimum qualifying score if you require a medical or criminal background waiver to enroll.

Programs for enlisting may be altered without prior notice by the service's current need for recruiting. The most recent information can be found by speaking with your recruiter.

The ASVAB lets you and the military assess your potential for several careers kinds if you are unsure about the type of work you want to undertake in the military.

Review all the chapters in this book, brushing up on the fundamentals of everything from science to electronics if you find yourself in this situation, but pay particular attention to the four subtests that will determine whether you are eligible for enlistment: Word

Knowledge, Paragraph Comprehension, Arithmetic Reasoning, and Mathematics Knowledge. Using this strategy, your suitability for different military occupations will be evaluated quite accurately.

Do-Over: Retaking the ASVAB:

No military branch will hire someone with an AFQT score between 0 and 9 since it indicates the person is not trainable. Even if your score is less than that, you cannot get the score you need to join the service branch of your choice. You must improve in at least one of the four main areas: word knowledge, paragraph comprehension, arithmetic reasoning, and mathematics knowledge.

When you're prepared, you can apply for the ASVAB (via your recruiter). After taking the ASVAB for the first time, you have one month to retake it (assuming the ASVAB in high school counts toward this requirement). You have to wait a month before testing after the initial retest.

You must wait at least six months before retaking the ASVAB after that. The ASVAB cannot be retaken on a whim or whenever you feel like it. In the following sections, I describe each service's specific policies about whether or not it permits retests.

As long as you are not serving in the military, the results of the ASVAB test are valid for two years. Your ASVAB scores are typically good as long as you are in the military after enlisting. In other words, barring a few circumstances, you can use your ASVAB enlistment test results to be eligible for retraining years later.

Army Retesting Policy in the United States:

The Army allows a retest in one of the following instances:

» The applicant's prior ASVAB test has expired.
» The applicant failed to receive an AFQT score high enough to qualify for enlisting.
» The applicant's previous ASVAB test has passed.
» Unusual situations arise, such as when a test taker cannot finish it for no fault.

Army recruiters aren't authorized to retest applicants to increase aptitude area scores to meet standards prescribed for enlistment options or programs.

Policy on retesting in the United States Air Force:

The purpose of retesting for the U.S. Air Force is for a candidate to boost their previous ASVAB scores to increase their alternatives for recruitment. The recruiting flight chief must conduct an in-person or phone interview with the candidate before approving the administration of any retests.

Remember the following additional regulations:

» After enrolling in the Delayed Entry Program (DEP), applicants are not permitted to retest by the Air Force.
» The current regulation allows candidates who do not hold a job/aptitude area reservation, which is not in the DEP but has already received qualifying test results, to retest.
» Retesting is permitted where the candidate's present line scores (mechanical, administrative, general, and electronic) make it difficult to link their qualifications to an Air Force competence.

Policy on retesting in the United States Navy:

The Navy allows the retesting of applicants:

» Who's previous ASVAB tests have expired.
» Who fails to achieve a qualifying AFQT score for enlistment in the Navy?

In most cases, Delayed Entry Program (DEP) individuals can't retest.

Policy on retesting in the United States Marine Corps:

The Marine Corps permits a retest if the applicant's previous test has expired. Otherwise, recruiters may ask for a retest if the initial results don't reflect the applicant's capability in light of the applicant's education, training, and experience.

The candidate can only request a retest for the Marine Corps if their original test results fall short of the requirements for their desired enlistment option or program.

Policy on retesting in the United States Coast Guards:

Before retaking the exam purely to improve scores to meet the requirements for a specific enlistment option, applicants for Coast Guard enlistments must wait six months after their last test.

After a month has passed since the initial ASVAB test, the Coast Guard Recruiting Center may approve retesting if there is good reason to think the results need to accurately represent the applicant's background in education, training, or experience.

Chapter 3
ASVAB Study Strategies

Setting Goals and Creating a Study Plan:

Setting Goals:

a. Understand the ASVAB Test Structure:
The ASVAB consists of nine subtests, including General Science (GS), Arithmetic Reasoning (AR), Word Knowledge (WK), Paragraph Comprehension (PC), Mathematics Knowledge (MK), Electronics Information (EI), Automotive and Shop Information (AS), Mechanical Comprehension (MC), and Assembling Objects (AO). Familiarize yourself with the content of each subtest and the time allotted for each section.

b. Identify Your Target Score:
Research the minimum qualifying scores for the branch of service you're interested in and identify your target score. Remember that higher scores can provide better job opportunities and enlistment bonuses.

c. Assess Your Strengths and Weaknesses:
Take an initial diagnostic ASVAB test to evaluate your current knowledge and skills. Analyze your performance and identify areas that need improvement.

d. Set Realistic, Achievable Goals:
Set SMART (Specific, Measurable, Achievable, Relevant, and Time-bound) goals based on your target score and initial assessment for your ASVAB test preparation.

Creating a Study Plan:

a. Allocate Study Time:
Determine your time until your test date and allocate adequate study time each week. Remember to balance your study sessions among all subtests, focusing extra on your weaker areas.

b. Gather Resources:
Collect up-to-date study materials, including official ASVAB test prep books, online resources, and mobile apps. Ensure to include a mix of content review, practice questions, and full-length practice tests.

c. Break Down Content:
Divide the content of each subtest into smaller, manageable topics. Create a study schedule that covers all issues and allows periodic reviews and practice tests.

d. Use Active Learning Strategies:
Employ effective learning techniques, such as summarizing information, creating flashcards, teaching others, and solving practice problems. These strategies promote better retention and understanding of the material.

e. Monitor Your Progress:
Regularly track your progress by taking practice tests and reviewing your performance. Adjust your study plan to address any persistent weaknesses or maintain your strengths.

f. Seek Support:
Join study groups or online forums, and contact mentors or tutors if you need additional help. Engaging with others can provide motivation, encouragement, and valuable insights.

    g.   <u>Develop Test-taking Strategies:</u>

Familiarize yourself with the ASVAB test format and develop strategies for managing time, eliminating wrong answers, and staying calm under pressure.

    h.   <u>Stay Consistent and Adapt:</u>

Stick to your study plan as much as possible but be flexible and adapt your approach if necessary. Consistency and perseverance are crucial to achieving your ASVAB goals.

Conclusion:

Setting clear goals and creating a personalized study plan are crucial steps to ensure success on the ASVAB test. Following the advanced information and strategies outlined in this guide will make you well-equipped to perform your best and achieve the scores necessary to pursue your desired military career.

Best Practices for Studying Effectively:

Studying effectively is crucial for academic success and long-term retention of information. This guide will discuss accurate and informative best practices for enhancing your study sessions and maximizing your learning potential.

Create a Conducive Learning Environment:

    a.   <u>Find a Quiet Space:</u>

Choose a quiet and distraction-free environment to help you focus during your study sessions.

    b.   <u>Organize Your Workspace:</u>

Keep your study area clutter-free and organized, with easy access to essential materials like textbooks, notebooks, and stationery.

    c.   <u>Minimize Distractions:</u>

Turn off or silence electronic devices and use website blockers or productivity apps to prevent access to distracting sites or notifications.

Employ Effective Time Management Techniques:

    a.   <u>Use a Study Schedule:</u>

Allocate specific time blocks for studying each day and plan your study sessions, considering deadlines and other commitments.

    b.   <u>Break Tasks into Smaller Chunks:</u>

Divide larger tasks into smaller, manageable segments using the Pomodoro Technique or other time management methods.

    c.   <u>Prioritize Your Tasks:</u>

Use the Eisenhower Matrix or other prioritization techniques to identify urgent and essential tasks and allocate your time accordingly.

Apply Active Learning Strategies:

    a.   <u>Summarize Information:</u>

Paraphrase and summarize content in your own words to ensure comprehension and retention.

    b.   <u>Teach Others:</u>

Explain concepts to a friend, family member, or peer; teaching others can help solidify your understanding.

    c.   <u>Use Visual Aids:</u>

Create mind maps, diagrams, or flowcharts to visualize complex concepts and relationships.

    d.   <u>Self-Test:</u>

Use flashcards, practice quizzes, or create your own tests to evaluate your understanding and identify areas for improvement.

Leverage Spaced Repetition and Interleaving:

a. <u>Spaced Repetition:</u>
Review information gradually to strengthen long-term memory retention.

b. <u>Interleaving:</u>
Mix different topics or subjects within a single study session to enhance cognitive flexibility and problem-solving skills.

Implement Metacognitive Strategies:

a. <u>Self-Reflection:</u>
Periodically assess your understanding, identify knowledge gaps, and adjust your study techniques as needed.

b. <u>Goal Setting:</u>
Set specific, measurable, achievable, relevant, and time-bound (SMART) goals for your study sessions and monitor your progress.

Optimize Your Mental and Physical Well-being:

a. <u>Get Adequate Sleep:</u>
Ensure you receive sufficient sleep, as sleep deprivation can negatively impact memory, focus, and overall cognitive performance.

b. <u>Exercise Regularly:</u>
Regular physical activity boosts cognitive function, reduces stress, and improves overall well-being.

c. <u>Maintain a Balanced Diet:</u>
Consume a well-balanced diet rich in nutrients, as proper nutrition can improve brain function and energy levels.

d. <u>Manage Stress:</u>
Use relaxation techniques like deep breathing exercises, meditation, or mindfulness to reduce stress and enhance focus.

Seek Help and Collaboration:

a. <u>Join Study Groups:</u>
Collaborate with peers to exchange ideas, clarify doubts, and gain new perspectives.

b. <u>Reach Out to Instructors:</u> Consult your teachers or professors for guidance, additional resources, or clarification on challenging topics.

Conclusion:

By incorporating these accurate and informative best practices for studying effectively, you can enhance the quality of your study sessions, improve your understanding of complex concepts, and ultimately achieve tremendous academic success. Remember to continuously evaluate and adjust your strategies to find the best approaches for your unique learning style.

Taking the Exam: Computerized or Paper?

There are numerous ASVAB variations, but you cannot control which one you take. The primary distinctions between the versions are between the electronic and paper versions.

Each version has benefits and drawbacks, which I follow in the following sections.

You will take the paper version of the ASVAB, which excludes the Assembling Objects subtest if you take it as part of the high school student program. You will take the enlistment ASVAB if you're taking the test to join any military branch. Two versions are available for this version: paper and pencil (P&P) and computerized (CAT-ASVAB).

The Pre-screening, Internet-delivered Computerized Adaptive Test (PiCAT) can be taken at your convenience. In any case, there is a high likelihood that you will take a computerized test because recruiters sometimes accompany their applicants to the closest

Military Entrance Processing Station (MEPS) for testing, medical examinations, and enlistment (one-stop shopping) to save time and money.

The P&P version is only offered in Military Entrance Test (MET) locations that are not conveniently located for MEPS, and MEPS exclusively uses the computerized version. If your high school plans a testing event, you should also take the P & P version. If you can't finish everything in one day, your recruiter can arrange an ASVAB-only test session and have you return for a follow-up physical (and to sign your contract).

Each state has a MET site (typically at a local high school or one of the 65 MEPS centers spread across the United States and Puerto Rico).

Note:

> You are kicked out of the testing facility if you cheat. However, even if you failed the test while peering at your neighbor's newspaper or computer screen, you would still fail. There are numerous test versions, and those seated nearby receive questions in various arrangements.

The Advantages and Disadvantages of Computerized Testing:

The computerized ASVAB uses adaptive testing to ensure that each candidate gets questions appropriate for their skill level. Therefore, the test questions are presented differently in this variation, known as the CAT-ASVAB. It is called adaptive because it adjusts the questions it presents to you based on your degree of skill.

The first exam question is of moderate difficulty. The following might be more challenging if you successfully respond to this question. If you answer incorrectly to the first inquiry, the machine will ask you an easy question. The easy and challenging questions are distributed randomly on the paper version of the ASVAB.

Additionally, the CAT-ASVAB features fewer questions than the paper-and-pencil version; this was done intentionally by the individuals who devised it. With this kind of examination, the computer can swiftly ascertain your level of knowledge without posing you with a wide range of questions that range from extremely easy to very difficult.

Note:

> Military recruiters have observed that many applicants who have taken both the paper-based and computerized versions of the ASVAB tend to score slightly higher on the computerized version of the test. This may be because people today are more at ease using a computer than a pencil.

You don't have to be a computer guru to appreciate the advantages of the computerized version of the ASVAB:

> You don't have to be a computer guru to enjoy the advantages of the computerized version of The ASVAB:
>
> » **It's impossible to record your answer in the wrong space on the answer sheet.** Questions and possible answers are presented on the screen, and you press the key corresponding to your answer choice before moving on to the next question. Only the A, B, C, and D keys are often activated when you take the test.
>
> » **The difficulty of the test items presented depends on whether you answered the previous question correctly.** On the two math subtests of the ASVAB, more complex questions are worth more points than more straightforward questions, so this method helps maximize your AFQT score.
>
> » **You get your scores right away.** The computer automatically calculates and prints your standard scores for each subtest and line scores for each service branch. (For more For line scores, see Chapter 2.) This machine is a smart cookie — it also calculates your AFQT percentile score on the spot. You usually know whether you qualify for the military. Enlistment on the same day you take the test and, if so, which jobs you qualify for.

The CAT-ASVAB has the drawback that you cannot skip questions or modify your responses once you have submitted them. You must respond to each question as it is asked rather than being able to go through and immediately respond to all the ones you are sure of.

This can make it challenging to choose how long to think about a tough subject before making a guess and moving on.

Additionally, even if you have a few extra seconds after the test, you can't double-check that you marked the correct response for every question.

Lastly, the CAT-ASVAB is the only test version that contains tryout questions (see Chapter 1 for more details), which can lengthen the overall amount of time you spend taking the test, but on the plus side, the tryout questions have no bearing on your final score.

The Benefits and Drawbacks of the Paper Version:

- The CAT-ASVAB questions are the same ones you get on the paper version. However, some people feel that the P&P ASVAB provides certain advantages:

» You can skip questions you don't know the answer to and return to them later.

This option can help when racing against the clock and wanting to get as many answers as possible. You can change an explanation on the subtest you're currently working on, but you can't change a response after the time for that subtest has expired.

» You may not make any marks in the exam booklet; however, you may make notes on your scratch paper.

If you skip a question, you can lightly circle the item number on your answer sheet to remind yourself to return to it. If you don't know the answer to a question, you can mentally cross off the answers that seem unlikely or wrong to you and then guess based on the remaining responses. Remember to erase any stray marks you make on your answer sheet before time is called for that subtest.

- The paper-based exam has additional drawbacks besides the destruction of trees. Additional negatives include:

» Harder questions are randomly intermingled with more straightforward questions.

This means you can spend too much time figuring out the answer to a question that's too hard for you and may miss answering more straightforward questions at the end of the subtest, lowering your overall score.

» The paper answer sheets are scored using an optical mark scanning machine.

The machine has a conniption when it comes across an incompletely filled-in answer circle or a stray pencil mark and will often stubbornly refuse to give you credit, even if you answered correctly.

» Getting your scores may seem like it takes forever.

The timeline varies; however, your recruiter will have access to your score no later than 72 hours (3 days) after you finish the test (not counting days MEPS personnel don't work, such as weekend days or holidays).

Dealing with Multiple-Choice Questions:

The computerized and paper versions of the ASVAB is multiple-choice tests. You choose the correct (or most correct) answer from the four available choices. Here are some tips to remember as you approach the choices:

» Read the directions carefully.

Most ASVAB test proctors agree that misreading recommendations is a prime offender when there's an issue with an applicant's scores.

Each subtest is preceded by a paragraph or two describing what the subtest covers and instructions on answering the questions.

» Make sure you understand the question.

If you don't understand the question, you're naturally not going to be able to make the best decision when selecting an answer. Understanding the question requires attention to three points:

- **Take special care to read the questions correctly.**

Most questions ask, "Which of the following equals 2 x 3?" But sometimes, a query may ask, "Which of the following does not equal 2 x 3?" You can skip right over the word, *not* when you're reading, assume that the answer is 6, and get the question wrong.

- **On the math subtests, be especially careful to read the symbols.**

When you're in a hurry, the + and ÷ signs can look very similar. Blowing right by a negative sign or another symbol is just as easy.

- **Make sure you understand the terms being used.**

When a math problem asks you to find the product of two numbers, be sure you know what finding the product means (you must multiply the two numbers). If you add the two numbers, you arrive at the wrong answer.

» **Take time to review all the answer options.**

You select the correct answer from only four options on all the subtests. On the ASVAB, you're supposed to choose the most accurate solution. (Now and then, you do the opposite and choose the least correct answer.) Sometimes, several answers are correct for the question, but only one is the best. If you don't stop to read and review all the answers, you may not choose the correct one. Or if you check all the answer options, you may realize you misread the question.

Often, a person reads a question, decides on the answer, glances at the answer options, chooses the option that agrees with their answer, marks it on the answer sheet, and then moves on. Although this approach usually works, it can sometimes lead you astray.

» **If you're taking a paper test, mark the answer carefully.**

A machine scores the paper based ASVAB answer sheets. You must mark the answer clearly, so the device knows which answer you selected. This means carefully filling in the space that represents the correct letter. You've done this a million times in school, but it's worth repeating: Don't use a check mark, don't circle the answer, and don't let your mark wander into the next space.

If you must erase, ensure all evidence of your prior choice is gone; otherwise, the grading machine may credit you with the wrong choice or disregard your correct answer and give you no credit.

Incorrectly marking the answer sheet (answering Question 11 on the line for Question 12, Question 12 on the line for Question 13—you get the idea) is a genuine possibility.

Be especially careful if you skip a question you will return to later.

Incorrectly marking the answers can cause a real headache. If you fail to get a qualifying score, you must wait one month before retaking the ASVAB. Your journey to military glory through ASVAB torment may not be over.

If your retest AFQT score increases by 20 points or more within six months of a previous test, you'll be required by MEPCOM regulation to take an additional ASVAB test, called a *confirmation test*. (Confirmation tests can be taken only at MEPS facilities.) So, if you're not careful, you'll take three ASVABs when all you needed to take was one. Sound fun?

Intelligent Guessing:

Sometimes speculating is acceptable on the ASVAB. However, you can benefit from guessing on the paper and pencil version due to how the test is graded. The point system is broken down as follows:

If you choose the correct answer, you get one point (or more, depending on how the question is weighted).

> » If you don't answer a question, you get nada.
> » If you guess on a question and get the question wrong, you get nada—no worrying about losing points or getting any penalty.

When you have the option to make educated predictions, do so. On the CAT-ASVAB, attempt to choose answers for every question, even if you need more time to finish a subtest. You will be punished if you don't finish a subtest; the algorithm will score the questions you didn't complete as though you answered them randomly.

Rarely does the ASVAB computer system need to penalize a test taker. Most people have enough time to finish the computerized test (or at least get very near to finishing it) because it typically takes around two hours to complete.

ASVAB questions include four alternative answers, so there is at least a 25% chance that you will guess correctly, allowing you to raise your score. You can frequently reduce your options to three or fewer replies by applying simple deduction because there is almost always at least one far from the correct response.

When eliminating answer options, bear the following in mind:

> » Don't eliminate an answer based on how frequently that answer comes up. For example, if Choice (B) has been the correct answer for the last five questions, don't assume it must be the wrong answer for the question you're on just because that would make it six in a row.
> » A solution that always has *all, everyone, never, none,* or *no one* is usually incorrect.
> » The longer the solution, the more likely it's correct. The test-makers must get all those qualifiers in there so you can't find an example to contradict the right answer.
> » If you see phrases like "in many cases" or "frequently," that's often a clue that the test-makers are trying to make the answer most correct.
> » If two choices are very similar in meaning, neither is probably the right choice. Conversely, if two answer options contradict each other, one is usually right.

If you must guess, never change the answer unless you're convinced that you're changing it to the correct answer. For example, you answered incorrectly only because you had sweat in your eyes and didn't read the choices properly.

The United States Air Force Senior NCO Academy conducted an in-depth study of several Air Force multiple-choice test results over several years. It found that when students changed answers on their answer sheets, they changed from a correct answer to a wrong answer more than 72 percent of the time!

In Parts 3, 4, and 5 chapters, you find more hints for making educated guesses specific to those topics. If you guess on multiple questions throughout the test, choosing the same answer for every guess is a smart way to go. For example, all your guesses could be Choice (B). This technique increases your chances of getting more answers correct. However, if you can eliminate Choice (B) as a wrong answer, then choose a different answer option for that question.

Preparing for the ASVAB through Study and Practice:

This book's practice exams are excellent study tools. Take one of the tests before starting your studies. Try to mimic the testing set by completing the exam simultaneously, keeping track of your time, and forbidding distractions.

As they say in the military, "Train as your fight." The ASVAB is the same way. Spend most of your preparation for the CAT-ASVAB using online exams. The written practice exams in this book should be your focus if you intend to take the ASVAB on paper and pencil.

Learn how long each subtest takes you to finish so you can estimate how much time you must spend making intelligent guesses. Check your responses after finishing each practice exam to determine where to improve.

When you study for the ASVAB, follow these study habits to make the most of your time:

> » **Focus on the subtests that matter to you.**

If you are interested in pursuing a career in electronics, the Electronics Information subtest should be at the top of your list to ace. Although you'll want to ensure all your line scores are good (in case your desired job isn't available or you want to retrain later in your career), focusing on your expertise in certain areas of interest makes you a more desirable candidate.

> » **Concentrate on subject areas that need improvement.**

It's human nature to find yourself spending your study time on subjects you are interested in or are good at. For example, if you're a whiz at fixing cars, don't waste your time studying for the Auto & Shop Information subtests. You're already going to ace that part of the test. On the other hand, if you had a hard time in math during high school, you need to spend extra time brushing up on your arithmetic skills.

> » **Be a loner.**

You may want to study with a partner now and then to brainstorm answers and quiz each other, but most of your studying should be done independently.

» **Try to reduce distractions.**
Always study in a well-lit, quiet area away from pets, loud music, and the TV.

» **Study in long blocks of time.**
Studying for an hour or two once or twice a day is much more effective than studying for 15 minutes six times a day.

» **Schedule your study times when you are rested.**
Don't try to cram studying into your schedule. Quality over quantity!

» **Keep study breaks short.**
A few minutes every hour is sufficient. Don't ignore breaks completely, however. Studies show that taking short breaks while you study improves how well you're able to remember information.

» **Practice the actual act of test-taking.**
Practice marking answers correctly on the answer key and time yourself to see how long it takes you to answer questions. The strategies provided can help you focus on what's most crucial, depending on your time before the test. Take the second practice test after studying a little more. Try to replicate the testing settings once more. Verify your responses. Examine your results in comparison to those from your first test.

How have you changed? If so, carry on with your current course of study. If not, think again about how you're studying or whether you're allotting enough time for it. You can get more study advice from a teacher or school counselor. With the subsequent tests, keep honing your skills.

Take the last practice test before the ASVAB a few weeks beforehand. Review any troublesome regions that are still giving you trouble. Examine your weaknesses to determine which subjects you should focus on more.

Note:

> Taking the final practice test is highly recommended as it can significantly improve your chances of success on the ASVAB. Taking the final practice test will help you identify areas where you may need to focus your studying and will help you become more familiar with the format and structure of the exam.
>
> By preparing well and putting forth your best effort, you can increase your likelihood of achieving a positive outcome.

Don't waste time memorizing the practice questions in this guide or any other ASVAB study guide. You won't see the same questions on ASVAB.

Use this guide and the sample tests for two purposes:

» **To determine the subject areas in which you need to improve:**
Use the tips and techniques and standard study materials (like high school textbooks) to improve your knowledge of each subject.

» **To familiarize yourself with the types of test questions and how they're presented on the test:**
Getting a good idea of all the subtests will improve your test-taking speed. You won't have to spend time figuring out how a question seems. Instead, you can spend your time answering the question.

Making Your Study Schedule:

The ASVAB can determine the path of your whole military career, provided you achieve the minimum AFQT score your branch demands. As I describe in Chapter 2, the military employs precisely calculated line scores to evaluate which positions within the branch you select will be a good fit for you. This means your study strategy must be customized for the job you want and the time you have until test day.

To start, figure out the minimum scores required for your desired position using Appendix A's tables.

It's recommended to phone a recruiter to find out the most recent requirements because minimum score standards can change anytime.

Having a backup job in mind is also a good idea if your top choice isn't available. Even if your recruiter is reasonably confident that your dream job will be available when you sign on the dotted line, I advise you to get used to adapting to the unexpected because you'll encounter it frequently during the length of your military career. This is because the military's demands change virtually daily.

Although your recruiter will contact MEPS to inquire about your availability to take the ASVAB, the decision to do so ultimately rests with you. Ask for more time if you need it so you can prepare. Studying at your own pace is crucial to acquire all the required knowledge. Additionally, avoid trying to put off studying until the night before the test because your performance will suffer due to being stressed on test day. Regardless of your timeline, the advice in the following section can assist you in developing a strategy.

Tip:

> "Make sure you give each subtest the necessary study time. The AFQT is one of the most important scores, but every section of the ASVAB is important. Each branch gauges your overall ability based on various combinations of your scores, so you need to do as well as possible on each subtest. Besides, scoring your best now means that if you want to switch jobs after you get some rank on your chest, you won't have to retake the ASVAB to score high enough to make the change."

Outlining Your ASVAB Study Strategy:

Plan your study strategy according to the subjects you must work on and how much time you have left. As I noted in the preceding section, if you only recently met with a recruiter (or haven't met with a recruiter at all) and you're still weighing your options, you can stretch out your timeline to give yourself plenty of time to prepare.

The following sections can help you map out a study plan if you know how much time you have to prepare for the test.

Ensure you're in a quiet place where others are unlikely to disturb you, and keep a stack of scratch paper and a pencil handy. Though you won't find the questions in this book (or in any other study guide) on the test, you will see the same concepts: knowing how to solve problems, decipher complex words, and dig up facts you may have learned long ago will help you perform your best on test day.

Take an objective look at your scores and determine which subtests you performed well on. Suppose they correspond with the job you want in the military. But as the military always says, don't become complacent. You must still study until the day before the test to ensure you're on the right track to ace the ASVAB.

You know exactly where to focus if you didn't perform well on some subtests. However, at this point, determining how you stack up against military standards on the AFQT portion of the test is most important.

The scores that matter most are:

- Word Knowledge
- Paragraph Comprehension
- Arithmetic Reasoning
- and Mathematics Knowledge.

These four subtests determine whether you're eligible to enlist in the military. If you performed poorly on one (or more) of them, head to Parts 2 (language arts) and 3 (math) to zero in on the topics you need. You may find creating your flashcards with blank index cards helpful. Studies have shown that flashcards help people memorize new information, and even writing down information can go a long way toward helping you remember.

Planning Your Review Three Weeks in Advance:

I understand you like to live on the edge. That's why you're taking the ASVAB three weeks from now and picking up this book. That, or your recruiter has told you that you're going to MEPS for testing in a few weeks without giving you much say. Either way, you're looking at about 21 days to tap-dance your way to the military job of your dreams.

First things first: Figure out what scores you need to qualify for that job (and a second choice, while you're at it) by flipping to Appendix A. Write down those scores so you have a goal you can look at as you study. Then, roll up your sleeves, grab a pencil, and take the first practice exam in Chapter 18. Score your answers as soon as you can (preferably right after you take the test) so you can identify your weak spots. If you take the practice test online, the software scores your answers.

Taking a full practice test immediately is essential. That way, if you have a blind spot—an area you're a little rusty in—you notice it and can start studying right away. If you take the first practice exam and realize you need more time to study before taking the ASVAB, that's okay—even if your recruiter wants you to move faster.

It's *your* career, not your recruiter's, and earning the scores you need to get a job you want pays off. If you don't reach the minimum scores you need for your dream job, the military will try to funnel you into a position it wants you to perform. That job may not match your interests, affecting your quality of life after signing up.

Tell your recruiter you need a few extra weeks (or more) to ensure you're ready; a good recruiter will understand. And don't forget that you aren't under any obligation to stick with the same recruiter. You can work with someone else in the office or even go to another Armed Forces Career Center. You can even talk to recruiters from other branches. Often, recruiters are direct reflections of the branches they serve.

Even if you score well on the subtests you need to ace for your dream job, you still need to head to those chapters in this book and study. Try to limit your study sessions to an hour or two a day to avoid burnout. Take the second one a week after you take the first practice exam. If you're very confident in some areas, skipping them is okay.

You must use your time wisely right now. That also means bypassing tests that don't contribute to your goals. For example, only the Navy uses scores from the Assembling Objects test, and even then, only for a handful of jobs, so practicing on it may be a waste of time if you aren't after one of those gigs.

Regardless of whether you skip specific subtests, score your answers right away. Then, look at the questions you answered incorrectly. They may help you pinpoint specific areas of knowledge you need to work on. For example, if you can't tell the difference between calipers and needle-nose pliers but the job you want requires decent shop knowledge, you probably need to study tools.

Before you complete your study, taking at least one AFQT practice exam is a good idea to ensure you have enough basic knowledge to qualify for enlistment. You should have a week before test day after taking the first and second practice exams and an AFQT practice exam. Take that time to revisit your most important topics and, if necessary, check out the additional resources I point out at the beginning of each chapter.

CHAPTER 4

ASVAB Test Day

What to Expect on Test Day?

On the day of your ASVAB test, knowing what to expect is essential to help you stay calm and focused. Here's a breakdown of what you can anticipate on test day:

a. **Arrival and Check-In**
Arrive at the testing center at least 30 minutes before your exam. Please bring a valid photo ID, as you'll need it for verification purposes. You'll likely have to store your personal belongings in a designated area, as they are prohibited in the testing room.

b. **Testing Environment**
The testing room will be quiet and well-lit, with individual stations for each test-taker. A proctor will be present to ensure that all test rules and regulations are followed.

c. **Test Format and Timing**
The ASVAB test consists of multiple-choice questions divided into different sections, including General Science, Arithmetic Reasoning, Word Knowledge, Paragraph Comprehension, Mathematics Knowledge, Electronics Information, Automotive and Shop Information, Mechanical Comprehension, and Assembling Objects. The test duration varies depending on whether you're taking the paper-and-pencil or computer-adaptive versions (CAT-ASVAB).

d. **Breaks**
You may have short gaps between sections, depending on the test format. Use this time to relax, stretch, and refocus for the next team.

Tips for Staying Calm and Focused:

Staying calm and focused is crucial for achieving a high score on the ASVAB test. Here are some tips to help you maintain your composure during the exam:

a. Get a good night's sleep the night before the test to ensure you're well-rested and alert.
b. Eat a balanced meal before the exam to maintain your energy levels.
c. Practice deep breathing exercises to help calm your nerves and reduce anxiety.
d. Maintain a positive attitude and believe in your ability to succeed.
e. Focus on the question and keep your mind from wandering to other topics.

Strategies for Answering Questions Efficiently

Developing effective test-taking strategies can help you answer questions more efficiently and improve your performance. Here are some tips for tackling the ASVAB test:

a. Read each question carefully to ensure you understand what is being asked.
b. Eliminate any incorrect answer choices to narrow down your options.
c. Use the process of elimination to increase your chances of choosing the correct answer.
d. If unsure of an answer, make an educated guess rather than leaving the question blank, as there is no penalty for guessing on the ASVAB.

e. Keep track of your time and pace yourself to ensure enough time to answer all the questions in each section.
f. If you're taking the CAT-ASVAB, remember that the difficulty of the questions adapts based on your previous responses. So, focus on answering each question to the best of your ability.

Chapter 5

General Science (GS)

PART I: LIFE SCIENCE:

Although nutritionists don't always agree about what constitutes a healthy diet, specific facts are clear. The human body requires a combination of protein, carbohydrates, fat, minerals, vitamins, and fiber. Proteins, carbohydrates, and fats (macronutrients) are necessary to provide energy. Minerals, vitamins (micronutrients), and fiber are essential to maintain proper bodily functions.

Proteins are necessary for the body's maintenance, growth, and repair. Animal proteins are contained in meat, fish, eggs, and cheese. Vegetable proteins are found in peas, beans, nuts, and grains.

Carbohydrates include both starches and sugars. They are significant sources of energy for the body. Starches are found in bread, cereal, rice, potatoes, and pasta. Sugars are found in fruits, cane sugar, beets, and processed foods.

Fats also provide energy for the body. There are three types of fats: saturated, monounsaturated, and polyunsaturated. Saturated fats can raise bad cholesterol (LDL), but mono- and polyunsaturated fats can decrease bad cholesterol levels. Diets high in saturated fat can lead to high cholesterol, which can cause heart disease or stroke. Sources of saturated fats include meats, shellfish, eggs, milk, and milk products. Sources of monounsaturated fats include olives and olive oil, almonds, cashews, Brazil nuts, and avocados. Sources of polyunsaturated fat include corn oil, flaxseed oil, pumpkin seed oil, safflower oil, soybean oil, and sunflower oil.

Fiber is an essential part of a healthy diet that provides bulk to help the large intestine carry away waste matter. Good sources of dietary fiber include leafy green vegetables, carrots, turnips, peas, beans, and potatoes, as well as raw and cooked fruits and whole-grain foods.

Water is also essential for survival. The body loses approximately four pints of water daily, which must be replenished. Most foods contain water, facilitating proper water maintenance, although it is still necessary to drink water as well! Insufficient water consumption leads to dehydration, which can cause muscle cramps, dizziness, and, if not remedied, even death.

Minerals in small quantities are needed for a balanced diet. Some necessary minerals are iron, zinc, calcium, magnesium, and sodium chloride (salt). Calcium is essential for building strong teeth and bones. Iron, on the other hand, is necessary for red blood cell development. Minerals play many different roles in developing and maintaining a healthy body.

Vitamins, such as vitamins C and D, are organic compounds necessary for various physiological processes, from bone hardness to healthy gums. Fruits and vegetables are rich sources of vitamins. Vitamin D is unique in that one of the best sources comes not from your diet but from the Sun. Exposure to sunlight allows your body to synthesize its own vitamin D. A lack of the proper amount of necessary nutrients in the diet can lead to deficiency diseases. One such disease is iron-deficiency anemia, which may cause weakness, dizziness, and headaches. It is especially common among children, young adults, and pregnant women who do not get enough iron in their diets. Another deficiency disease is scurvy, caused by a lack of vitamin C. Though at one time very common among pirates and sailors who did not have access to fresh fruits and vegetables, scurvy is now relatively rare.

Human Body Systems and Diseases
THE SKELETON AND MUSCLES

The skeleton and muscles hold the body together for movement. Without a skeleton, you would be just an immobile mass of organs, veins, and skin. Some organisms, namely **arthropods** (insects, spiders, and crustaceans), have **exoskeletons** or external skeletons. However, **vertebrate** animals, including humans, have internal skeletons or **endoskeletons**.

The human skeleton contains both **bone** and **cartilage**. Bones provide primary support, while more flexible cartilage is found at the end of all bones, joints, nose, and ears. Bones provide structural support for the body, protect vital organs, produce blood cells, and store minerals such as calcium. **Tendons**, tough fibrous connective tissue cords, connect muscles to the skeleton. **Ligaments**, another type of connective tissue, connect bones to other bones at joints such as the elbow, knee, fingers, and vertebral column.

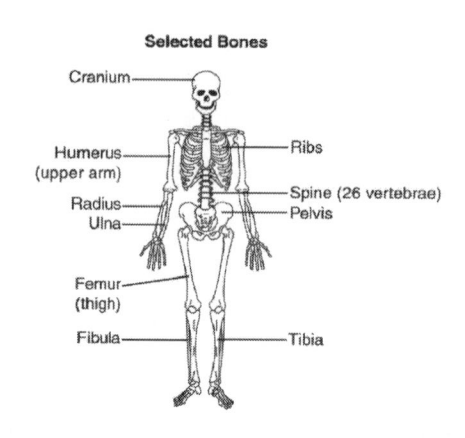

THE RESPIRATORY SYSTEM:

Respiration—the process by which blood cells absorb **oxygen** and eliminate **carbon dioxide** and **water vapor**—is performed by the **respiratory system**.

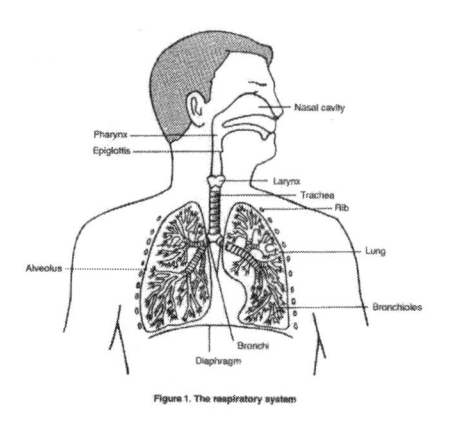

When air enters through the **nose**, it passes through the **nasal cavity**, which filters, moistens, and warms air, and then through the **pharynx**, which further purifies the air and aids in protection against infection. The air then passes through the open **epiglottis**, which closes when swallowing to prevent food from going down the airway, and into the **trachea**, which further cleanses the air.

The trachea branches into the **left and right bronchi**, two tubes leading to the **lungs**. There, the bronchi are further subdivided into smaller tubes called **bronchioles**. Each bronchiole ends in a small sac called an **alveolus**. In the alveolus, oxygen from the air enters the bloodstream via tiny blood vessels called **capillaries**.

The **diaphragm** is a system of muscles that allows breathing. When the diaphragm causes the lungs to expand, air rushes in to fill the space in a process called inhalation. When the diaphragm causes the lungs to contract, the air is pushed out in an exhalation.

BLOOD AND THE CIRCULATORY SYSTEM:

In conjunction with the respiratory system, the **circulatory system** transports oxygen throughout the body while removing carbon dioxide. Additionally, the circulatory system transports nutrients the digestive system provides and clears away waste by transporting it to the excretory system. The organ that drives the circulatory system is the **heart**.

Human nature is a four-chambered pump, with two collecting chambers called **atria** (singular: atrium) and two pumping chambers called **ventricles**. The **right atrium** receives deoxygenated blood from the **venae cavae** (plural of **vena cava**), the two most extensive veins in the body. It passes to the **right ventricle**, which pumps the blood to the lungs through the **pulmonary artery**. Blood picks up oxygen in the lungs and returns to the **left atrium** via the **pulmonary vein**.

From there, it passes to the **left ventricle** and is pumped through the **aorta**, the body's largest artery, into several smaller branching **arteries** that take it through the rest of the body. The heart's **valves** are essential to the efficient pumping of the heart. When blood is pumped out of the ventricles, valves close to prevent the blood from flowing backward into the heart after the contraction of the ventricles is complete.

> **How blood flows from the heart to the body:**
>
> right atrium → right ventricle → lungs → left atrium → left ventricle → body

The right side of the heart is associated with deoxygenated blood (because the blood hasn't gotten to the lungs yet), whereas blood coming into the left side of the heart is oxygenated because it's sent there from the lungs. **Heart disease (cardiovascular disease)** is the most common cause of death in the United States. High cholesterol, high blood pressure, smoking, and lack of exercise can all contribute to heart disease, leading to heart attack or heart failure.

The Human Heart:

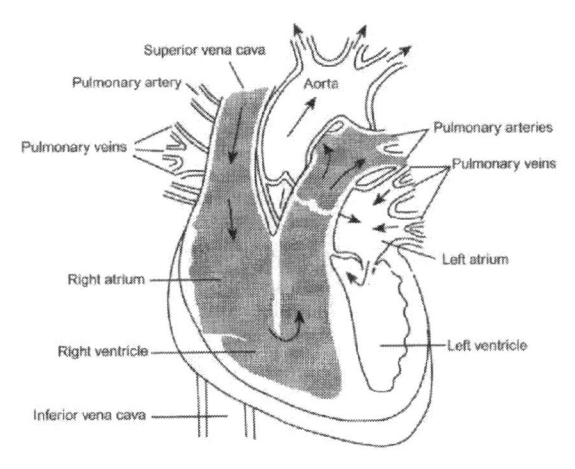

Figure 2. Human heart

The **arteries** carry blood from the heart to the body's tissues. They repeatedly branch into smaller arteries (**arterioles**), which supply blood to bodily tissues via the capillaries. Arteries carry blood away from the heart and thus must be thick-walled because they have oxygenated blood at high blood pressure. Only the pulmonary artery, which carries blood from the heart into the lungs, does not contain oxygenated blood.

Conversely, veins have blood back to the heart from other body parts. Veins are relatively thin walled, conduct blood at low pressure, and contain many valves to prevent backflow. Veins have no pulse and carry dark red, deoxygenated blood. The lone exception is the pulmonary vein, which carries freshly oxygenated blood from the lungs into the Heart.

Finally, **capillaries** are thin-walled vessels that are very small in diameter. Capillaries, rather than arteries or veins, permit the exchange of materials such as oxygen, carbon dioxide, nutrients, and waste between the blood and the body's cells through **diffusion**.

The Capillary System:

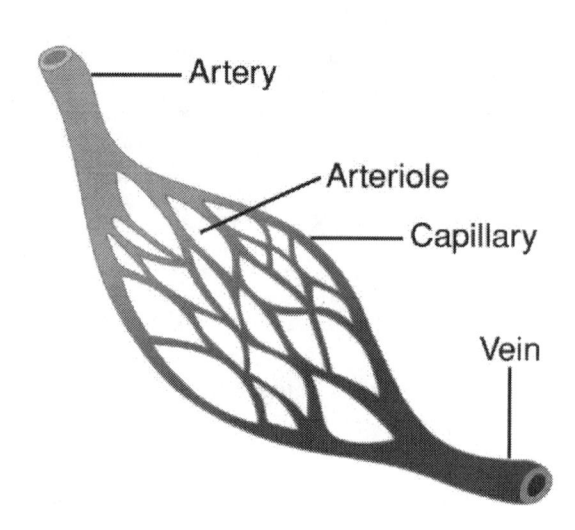

Hypertension, also known as high blood pressure, can cause damage to blood vessels as well as other parts of the body, like the kidneys. Limiting salt intake, maintaining a healthy weight, and exercising can help prevent or manage hypertension.

Blood consists of **cells** suspended in **plasma**, the liquid component of blood. Three types of cells are found in the blood: **red blood cells**, the oxygen-carrying cells; **white blood cells**, which fight infection by destroying foreign organisms; and **platelets**, which are cell fragments that allow blood to clot. All blood cells are created in the **bone marrow**, located in the center of bones.

These blood cell types can be measured to indicate overall health. When white blood cell levels are higher than usual, the body fights off some infection by either bacteria or a virus. It is also important to note that blood comes in four types: A, B, AB, and O, which can be further designated as negative or positive.

One combination may be written as A+ or A positive, for example. The type of molecules (antigens) found outside the red blood cells determines the letter designation. The positive or negative title is assigned based on whether cells have a third type of antigen called the **Rh factor**.

A person with blood that is Rh-factor negative cannot receive blood with a positive kind; however, a person with positive type blood can receive Rh-negative donor blood. Type O negative is the **universal donor**, which means that type O negative blood can be given to anybody. Type AB positive is the **universal recipient**, which means that someone with this type of blood can receive any other kind of blood.

THE DIGESTIVE AND EXCRETORY SYSTEMS:

The digestive system breaks down foods into materials the body can use for energy and building body tissues. The digestive tract is a long and winding tube that begins at the mouth and ends at the anus. The process of digestion progresses as follows:

In the mouth, the teeth, and the tongue aid in mechanical digestion (chewing), while the salivary enzyme **amylase** in the saliva begins to break down starch. From the mouth, the chewed food moves into the **esophagus**. Contractions push the food down through the esophagus and into the stomach.

In the **stomach**, food is mixed with **gastric acids** and **pepsin**, which help break down **protein**. Most digestion takes place in the **small intestine**. The small intestine is very long, about 23 feet on average. Food is broken down completely by enzymes produced in the walls of the small intestine, the pancreas, and the liver.

The acids produced by the **pancreas** contain **lipase**, which converts fat to glycerol and fatty acids; **pancreatic amylase**, which breaks down complex carbohydrates into simple sugars; and **trypsin**, which converts polypeptides (the molecules that compose proteins) into amino acids. **Bile**, which is produced by the **liver**, aids in digestion by emulsifying fat (physically separating it into individual molecules).

All these digested substances, except for the fatty acids and glycerol, are absorbed in the small intestine through capillaries that carry the blood into the liver and throughout the rest of the body.

In the **large intestine**, also known as the **colon**, water, and minerals remaining in the waste matter are absorbed back into the body. Chemical waste, such as urea, excess salts, minerals, and water, is filtered from the blood by the **kidneys** and secreted into the urine. Urine is transported to the bladder from the kidneys through the **Ureters**.

Solid waste matter is stored in the rectum. Liquid waste (urine) is stored in the **bladder**. Solid waste matter is periodically released through the **anus**, and urine is released through the **urethra**.

The Digestive System:

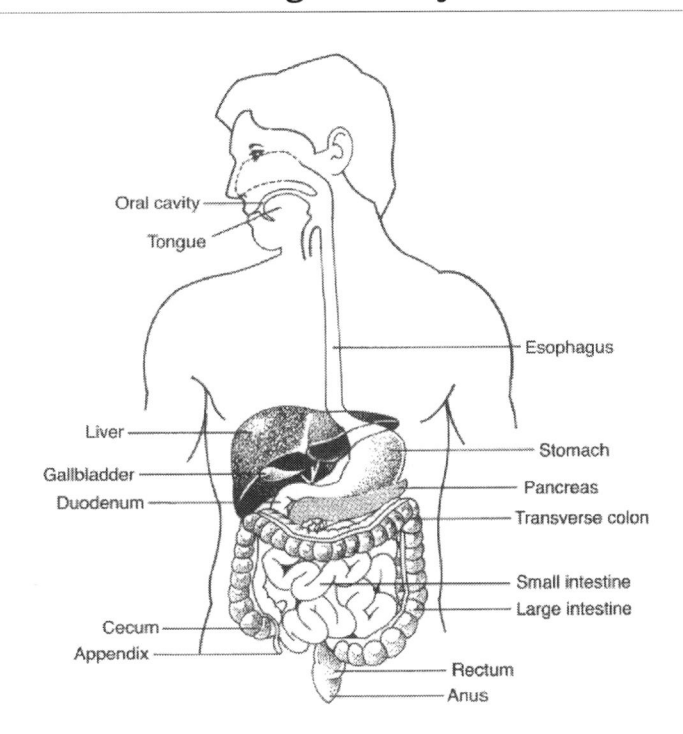

THE NERVOUS SYSTEM

The nervous system consists of the **brain**, the **spinal cord**, and the network of billions of nerve cells called **neurons**, which behave like electrical wires that send and receive signals throughout the body. The nervous system controls the body's functions and receives and processes environmental stimuli.

The nervous system comprises the **central nervous system**, the brain and spinal cord, and the **peripheral nervous system**, which contains all the other neurons throughout the body. The main components of the central nervous system are as follows: The **cerebrum** is the central part of the brain. It is thought to be the **center of intelligence**, responsible for hearing, seeing, thinking, etc.

The **cerebellum** is a big cluster of nerve tissue that forms the basis for the brain. It is most closely associated with balance, movement, and muscle coordination. The **medulla**, part of the brainstem, is the connection between the brain and the spinal cord. It controls **involuntary actions** such as breathing, swallowing, and heart beating. The **spinal cord** is the major connecting center between the brain and the network of nerves. It **carries impulses** between all organs and the brain and is also the control center for many **simple reflexes**.

The Human Brain:

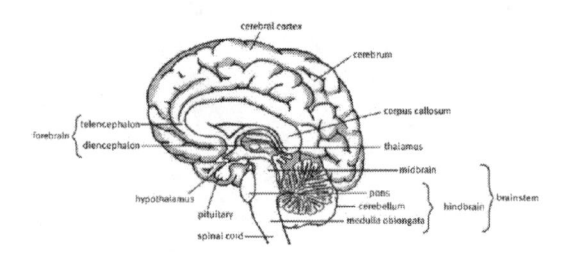

The peripheral nervous system can be subdivided into:

- The **somatic nervous system** consists of nerve fibers that send sensory information to the central nervous system and control **voluntary actions**.
- The **autonomic nervous system** regulates **involuntary heart, stomach, and intestine activity**.

THE REPRODUCTIVE SYSTEM:

Human reproduction occurs when a male's sperm fertilizes a female's egg. During female **ovulation**, which occurs approximately every 28 days, an egg (**ovum**) is released from one of the ovaries and travels through the **oviduct (fallopian tube)** and into the **uterus**. At the same time, the **endometrial lining** of the uterus becomes prepared for implantation.

Male and Female Reproductive Systems:

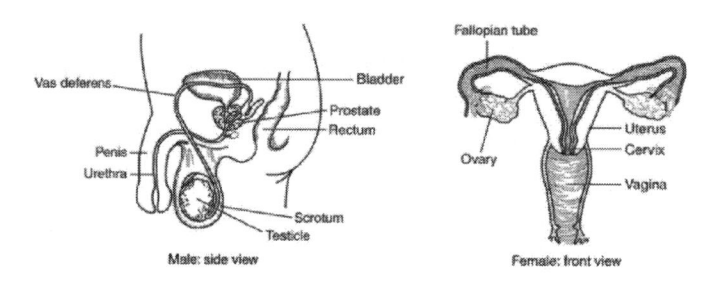

During intercourse, the **penis** ejaculates more than 250 million **sperm**, produced in the **testes**, into the **vagina**. Some sperm make their way to the uterus, where they may encounter an egg to fertilize. If the sperm unites with the ovum, a fertilized egg (**zygote**) is formed, which may implant in the uterus and eventually develop into a **fetus**. During pregnancy and after childbirth, **prolactin**, a hormone secreted by the pituitary, activates breast milk production (**lactation**).

The uterine lining sloughs off during menstruation if the ovum fails to fertilize. From puberty to menopause, this **menstrual cycle** repeats monthly except during pregnancy.

HUMAN PATHOGENS:

Some diseases, such as deficiency disorders, hypertension, and heart disease, are caused by diet and/or lifestyle factors. However, another major cause of human disease is **pathogens**—disease-causing agents—such as bacteria and viruses.

Bacteria, single-celled organisms, are responsible for conditions such as strep throat, staph infections, and pneumonia; these illnesses may be treated with antibiotic medications. On the other hand, **viruses** are not technically living things because they can only replicate inside a host's cells.

Viral human illnesses include the common cold and flu, AIDS (acquired immunodeficiency syndrome), and herpes. These illnesses cannot be treated with antibiotics but may be treated with specially designed antiviral drugs. Both bacteria and viruses may be spread from person to person in several different ways. Some bacteria and viruses (including the cold and flu) may be passed through the air, wherein an infected person coughs or sneezes and another individual inhales the pathogen. Some viruses, like HIV (human immunodeficiency virus), can only be transmitted through contact with infected body fluids, as in sexual intercourse or intravenous drugs. Others, like herpes, can be spread by skin-to-skin contact. Every virus and bacteria have a unique profile regarding how it is transmitted and what kind of cells it infects.

Luckily, many diseases caused by viruses (and a few caused by bacteria) can be prevented through **vaccination**. Vaccination, also called **immunization**, prevents many diseases that, not long ago, would have been very severe if not fatal, including smallpox, polio, and measles.

When a person receives a vaccine, a small amount of deactivated, weakened, or partial pathogen is injected into the body, causing the immune system to react; if the body is exposed to that pathogen in the future, the immune system will have a quick response to it, protecting the person from infection.

Genetics:

Genetics is the study of heredity, the process by which characteristics are passed from parents to offspring. The basic laws of genetics have been understood since the late eighteenth century when Gregor Mendel first discovered them.

Mendel discovered that in sexual reproduction, individual heredity traits separate in the reproductive cells so that reproductive cells, known as **gametes**, have half as many **chromosomes** (large strings of hereditary units) as normal cells. Normal body cells are called **diploid**, and gametes are called **haploid**. (Think *DIploid = Double* and *HAploid = HAlf* to help you remember it.)

In human reproduction, the female gamete (ovum) combines with the male gamete (sperm), each of which contains 23 unpaired chromosomes(haploid), to produce a zygote, which contains 23 pairs of chromosomes, or a total of 46 (diploid). **Meiosis** is the process by which gametes are created—sexual reproduction by meiosis and fertilization results in many variations among offspring.

DNA (deoxyribonucleic acid) is a molecule that contains genetic information. It is "written" with the **genetic code**, a combination of **nucleotides** that bind together in a specific pattern that can be "read" by the cell to instruct how to grow and behave. DNA is shaped in a **double helix**, a twisted ladder along its axis.

DNA Molecule

A **gene** is defined as the unit of inheritance contained within an individual's DNA. A gene may come in several varieties, known as **alleles**. For example, a person may have the allele for brown or green eyes in the gene that determines eye color.

Each of us has two alleles for every gene, one inherited from each parent. These alleles may or may not be alike. If the alleles are alike, that person is **homozygous** for that gene. If the alleles differ, the person is **heterozygous** for that specific gene.

The genes on the sex chromosome determine the sex of babies. In females, the two sex chromosomes are alike and are designated as XX. In males, the sex chromosomes are heterozygous and designated as XY.

Genetic traits are inherited independently of one another. When different traits are paired up during an egg's fertilization, one trait is often **dominant,** and the other is **recessive**. A recessive trait is only expressed if the offspring has two copies of that trait. A dominant trait, on the other hand, will be expressed even if only one copy (paired with a recessive trait) is present.

For example, Huntington's disease, which causes degeneration of nerve cells and loss of muscle control, is passed from parent to offspring as a dominant trait, so only one copy is necessary for the offspring to inherit the disorder. On the other hand, cystic fibrosis, which affects the lungs, is caused by a recessive trait, which must be inherited from both parents to cause the disease.

Other disorders, such as some types of cancer, Down syndrome, and color blindness, are also caused by genetic traits.

A person's **genotype** is their genetic makeup, including dominant and recessive alleles. A person's **phenotype** is how their genes express themselves in physical characteristics. Take eye color, for example. If a woman has brown eyes, then brown eyes are part of her phenotype.

Cellular Structures and Functions:

Cell theory states that (1) all living things are composed of cells, (2) cells are the basic units and structure of living things, and (3) new cells are produced from existing cells. Cells are classified into two categories based on the absence or presence of a nucleus: **prokaryotic** and **eukaryotic**. Prokaryotic cells are characterized by not having a nucleus; bacteria are one example.

Structure of a Prokaryotic Cell:

Plants, animals, fungi, and protists comprise eukaryotic cells with a nucleus and a more complex structure than prokaryotic cells. The **nucleus** of a eukaryotic cell contains the genetic material of the cell. Outside the nucleus lies the **cytoplasm**, which surrounds the other cell structures.

The cytoplasm contains many other **organelles** (cell parts with specific functions).

These include:

- **Ribosomes**, which produce proteins.
- **Mitochondria** (singular: *mitochondrion*), which produce energy.
- **The endoplasmic reticulum** is involved in the synthesis of proteins and fats.
- **Golgi apparatus**, which "packages" proteins for use.
- **Lysosomes** help the cell manage waste.
- **Centrosomes**, which can be crucial in guiding the cell's reproduction.

Structure of a Eukaryotic Cell:

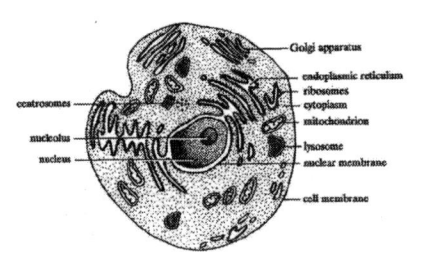

Plant cells also have a somewhat rigid **cell wall** surrounding the membrane. The cell wall provides structure and support for cells. Some plant cells produce their energy through the process of photosynthesis. **Photosynthesis** is when sunlight, carbon dioxide, and water react to make sugar and oxygen. It serves as a source of energy for the cells and occurs in plants.

Animal cells are surrounded by a semipermeable membrane, which allows for the diffusion of water and oxygen from inside the cell to the outside of the cell and vice versa. These cells cannot produce their energy and rely on consuming outside sources to provide them with the tools to make energy through cellular respiration.

Cellular respiration is when the mitochondria process sugar and oxygen to produce energy, water, and carbon dioxide. Cellular respiration serves as the energy source for animal cells. If no oxygen is present, cellular respiration will result in fermentation, where either lactic acid or alcohol is produced instead of sugar.

Cell division is the process where genetic material is replicated in the nucleus. Cell division begins in interphase, where DNA replication occurs. This results in the replication of the chromosomes in the nucleus. **Chromosomes** are tightly coiled threads of DNA composed of twin strands called **chromatids**.

Interphase is the longest part of cell division, divided into cell growth and DNA replication periods. The cell grows to accommodate the increase in chromosomes.

The following interphase is the prophase. During prophase, chromatids begin to pair up with their sister chromatids.

This leads to metaphase, where the sister chromatids move to opposite cell poles.

During the next phase, anaphase, the chromatids begin to pull apart into two separate poles. The cell becomes elongated during this phase, which makes it very easy to identify.

During telophase, the two new nuclei become completely separated.

The final phase in cell division is cytokinesis, where the cytoplasm and cell membranes complete their separation, and two daughter cells are formed.

Typically, cell reproduction is closely regulated by genetic signals in the cell that tell it when to stop reproducing; **cancer** occurs when the signals are mutated, and cells can grow without limit. Factors such as smoking, sun exposure, and genetic mutations can cause damage to cells and may lead to cancer. The most common type of cancer is skin cancer, caused by exposure to UV rays in sunlight.

Ecology:

Ecology is the study of the interrelationships between organisms and their physical surroundings. Just as biologists classify organisms according to terminology that goes from the general to the specific, ecologists employ a similar set of terminology:

- **Biosphere**: The zone of planet Earth where life naturally occurs, including land, water, and air, extending from the deep crust to the lower atmosphere.
- **Biome:** A major life zone of interrelated species bound by similar climate, vegetation, and animal life.
- **Ecosystem**: A system comprising a community of animals, plants, and other organisms, as well as the abiotic (non-living) aspects of its Environment. An ecosystem, including pristine and highly developed areas, can be large or small. An ecosystem contains a community, and this community may contain many populations of organisms. The various populations within a community fall into one of several roles in the food Chain.
- **Community**: The collection of all ecologically connected species in an area.
- **Population**: A group of organisms of the same species living in the same region. **Producers** (mainly plant life): Also known as **autotrophs**, they make their food via photosynthesis.
- **Decomposers** (bacteria and fungi): Also known as **saprotrophs**, they break down organic matter and release minerals into the soil.
- **Scavengers** (many insects and certain vertebrates, such as vultures and jackals): These animals exhibit characteristics of decomposers by consuming refuse and decaying organic matter, especially **carrion** or decaying flesh. Similar organisms called **detritivores** consume small pieces of decaying organic matter called detritus that are too small for most scavengers to want. **Consumers** (most animals): Also known as **heterotrophs**, they consume other organisms to survive. Consumers are divided into three types:
- **Primary consumers**: Also known as **herbivores**, they subsist on produce, such as plants. Examples include grasshoppers, deer, cows, and rabbits.
- **Secondary consumers**: Also known as **carnivores** or predators, subsist mainly on primary consumers. Examples of secondary consumers include birds of prey (such as owls and falcons), foxes, and snakes. Some secondary consumers are also **omnivores**, meaning they consume producers and consumers. Examples include chickens, rats, lizards, and sea otters.
- **Tertiary consumers**: They can eat secondary consumers, also known as top carnivores. Many tertiary consumers are also **omnivores**. Examples include lions, wolves, sharks, and human beings.

The diagram below shows roughly the relationships between these groups in an ecosystem:

Hierarchy of Consumers:

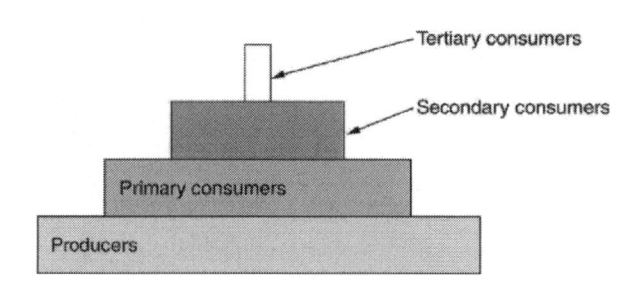

Food chains are a basic way to see the levels in an ecosystem. However, food webs can show more complex relationships among the levels. In the example below, the organism to which an arrow points might eat the organism on the other end of that arrow.

A Food Web

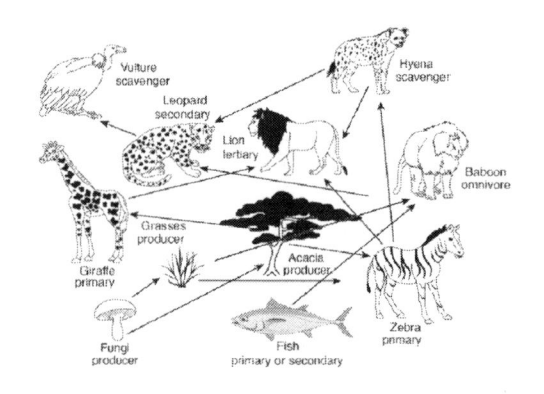

Study this example of a solid approach to an ecology question.

The Classification of Living Things:

All living things fall into a careful classification scheme that goes from the broadest level of similarity (**domain**) to the narrowest (**species**). The classification scheme has seven different levels. They are **Domain**: the broadest category; there are only three domains. This level of classification is a relatively recent change to the classification system. You may not have been taught about domains in school, and you may or may not see them on the ASVAB.

- **Kingdom**: which contains several related *phyla*.
- **Phylum**: which contains several related *classes*.
- **Class**: which contains several related *orders*.
- **Order**: which contains several related *families*.
- **Family**: which contains several related *genera*.
- **Genus**: which contains several related *species*.
- **Species**: which contain organisms so similar that they can only reproduce with one another to create viable, fertile offspring.

For example, this is the classification of human beings:

- Domain — Eukaryota
- Kingdom — Animalia
- Phylum — Chordata
- Class — Mammalia
- Order — Primates

- Family — Hominidae
- Genus — Homo
- Species — Sapiens

The three domains are:

Eukaryota: All living things whose cells have nuclei are in this domain. Almost all multi-celled organisms (including plants, animals, and fungi) are in this domain.

Bacteria and **Archaea:** These domains contain single-celled organisms whose cells do not have nuclei. Living things in the two domains are distinguishable by metabolic and chemical differences.

Classes, orders, families, genera (plural of genus), and species are among the numerous phyla (plural of phylum). You may use online resources to learn about some of the most significant phyla, classes, etc. It is essential to achieve your professional objectives if you have extra time before Test Day, have passed all the other subject exams, and have increased your General Science score. Otherwise, it's unlikely that studying will be the best use of your time.

Life Science Practice Questions:

1. Which of the following describes the proper pathway of blood through the heart?

(A) vena cava → right atrium → right ventricle → pulmonary artery → pulmonary vein → left atrium → left ventricle → aorta
(B) vena cava → right atrium → right ventricle → pulmonary vein → pulmonary artery → left atrium → left ventricle → aorta
(C) vena cava → right atrium → left atrium → pulmonary artery → pulmonary vein → left atrium → left ventricle → aorta
(D) vena cava → right atrium → left atrium → pulmonary vein → pulmonary artery → left atrium → left ventricle → aorta

2. Most human digestion takes place in the:
(A) esophagus
(B) stomach
(C) small intestine
(D) large intestine

3. Which blood type can be donated to anyone?
(A) A positive
(B) B negative
(C) O negative
(D) AB positive

4. A typical human gamete contains:
(A) 2 chromosomes
(B) 23 chromosomes
(C) 46 chromosomes
(D) 92 chromosomes

5. Which of the following is an example of a primary consumer?
(A) moss
(B) mushroom
(C) jackal
(D) deer

6. How many domains are recognized in taxonomy?
(A) 3
(B) 5
(C) 7
(D) 9

7. Which of the following foods is a good source of fiber?
(A) oils
(B) fruits and vegetables
(C) dairy products
(D) meat and seafood

8. Which components of a prokaryotic cell produce proteins?
(A) ribosomes
(B) mitochondria
(C) lysosomes
(D) centrosomes

9. A zygote is a:
(A) diseased cell
(B) mutated male reproductive cell
(C) female reproductive cell
(D) fertilized egg

10. Which of the following describes how the immunization process takes place?
(A) A highly active form of a pathogen is injected, causing the formation of strong antibodies that will fight off the injected pathogen and remain in the bloodstream to fight off future infections.
(B) Synthetic antibodies are injected into the bloodstream that will persist in the body and fight off future infections.
(C) A small amount of a deactivated pathogen is injected, causing the immune system to react so that if an active form of the pathogen is encountered in the future, the immune system will respond quickly.
(D) Chemicals that can destroy a pathogen's reproduction ability are injected into the bloodstream.

Answers and Explanations
LIFE SCIENCE PRACTICE QUESTIONS:

1. A
To predict this question, think about what kind of blood (oxygenated or deoxygenated) flows into and out of the heart.

2. C
The small intestine is the largest digestive organ and does the most work breaking down food into materials the body can use.

3. C
Type O negative blood, also known as the "universal donor" type, can be donated to anyone.

4. B
A typical human gamete contains half the number of chromosomes as a normal cell, or 23.

5. D
A deer is an example of a primary consumer, an animal that consumes only vegetation.

6. A
There are three domains in taxonomy: Eukaryota, Bacteria, and Archaea.

7. B
Sources of dietary fiber include fruits and vegetables, whole grains, and legumes.

8. A
The primary function of ribosomes is to produce proteins. Mitochondria produce energy, lysosomes help a cell manage waste, and centrosomes are essential in a cell's reproduction.

9. D
When a sperm unites with an ovum, the product is a fertilized egg called a zygote.

10. C
Immunization uses a deactivated form of a bacterial or viral pathogen. Though inactive, this pathogen triggers the formation of antibodies that attack it, thus "training" the immune system to respond quickly should the active pathogen be introduced.

Part II
EARTH AND SPACE SCIENCE:

Earth and space science is the study of the Earth and the universe around it. For purposes of the ASVAB, it's helpful to know a few facts about our planet and the solar system in which it travels.

GEOLOGY:

Geology is the science that deals with the history and composition of the Earth and its life, especially as recorded in rocks.

STRUCTURE OF EARTH:

Scientists have discovered that the Earth comprises three layers through examining rocks. Roughly one percent of the Earth's total volume comprises the crust, the topmost layer. Its thickness ranges from 10 kilometers to 100 kilometers. The mantle, more than 75% of the Earth's bulk, lies beneath the crust.

The mantle is around 3,000 km deep, composed mainly of iron, magnesium, and calcium. It is significantly hotter and denser than the surface of the Earth because pressure and temperature inside the Earth rise with depth. The Earth's core is located at its center and is approximately twice as dense as the mantle due to its metallic (iron-nickel alloy) rather than stony makeup. The Earth's liquid outer core measures 2,200 kilometers thick, whereas the solid inner core has a radius of 1,300 kilometers.

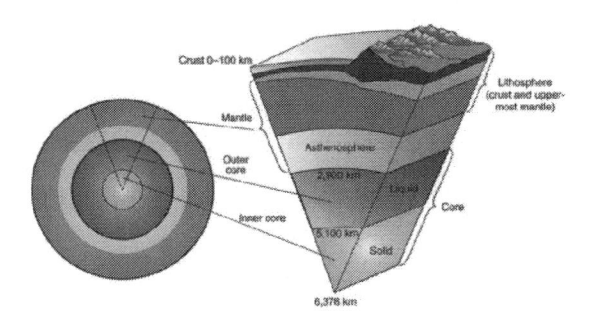

PLATE TECTONICS:

As previously mentioned, the Earth's interior is warm, ranging from 3,000°C to 4,000°C. The Earth's upper mantle and crust are solid rock, which generally prevents this heat from escaping. Roughly 30 plates comprise the crust and the hard upper mantle (the lithosphere).

Throughout hundreds of millions of years, these plates' gradual motion over the more mobile mantle beneath (the asthenosphere) has caused the continents to drift apart progressively. Fault lines run along the margins of these plates. (Places where the plates slide in relation to one another are called fault lines.)

Earthquakes can happen along fault lines when plates shift in relation to one another. Scientists gauge the strength of an earthquake using either the Richter scale or the moment magnitude scale. Perhaps more familiar to most people is the Richter scale.

The Richter scale starts at 1 and goes up in magnitude by an amount roughly 10 times greater than the step before.

GEOLOGIC TIME SCALE:

The majority of information on the part of our planet has been gleaned from fossil records preserved in sedimentary rock. By examining rocks, we now know that the Earth is roughly 4.6 billion years old and that for most of that time, very little evidence of fossils was left behind. Before the fossil record started, Precambrian eon describes the period between 4.6 billion years and 570 million years ago. However, life first evolved on Earth as early as 3.5 billion years ago, even though early geologists who examined the Precambrian eon could not identify early, rudimentary fossils!

Cycles in Earth Science:

Numerous biogeochemical cycles exist. They all function similarly in that an element or compound is discharged into the air and brought back to the earth. However, the water and carbon cycles are the two fundamental cycles on which this section will concentrate. The hydrologic cycle is another name for the water cycle. All forms of water, including solid, liquid, and gas, must flow through the atmosphere and return to the Earth.

As water evaporates (becomes a gas) from the ocean's surface and that of other bodies of water, it rises into the sky. Transpiration is how water vapor from plant leaves enters the atmosphere. Clouds are created when this water vapor condenses. When the clouds are too dense, they produce precipitation as a liquid (rain) or a solid (snow and ice). Some of the water that collects on the surface will descend as snowmelt or surface runoff and eventually return to the ocean via rivers and streams.

The remaining water undergoes a process called infiltration, where it is absorbed into the Earth's surface and moves to the water table. When wells are dug, the water table serves as a reservoir and can be drawn upon through the digging of wells.

The Water Cycle:

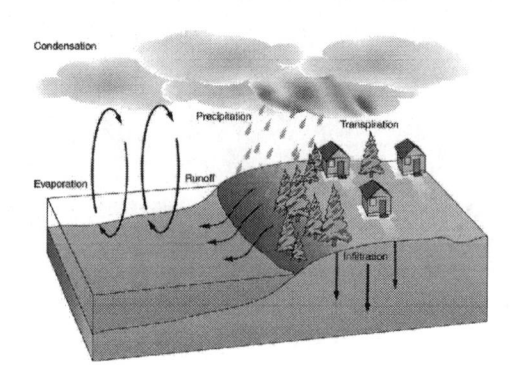

Carbon Cycle:

One of the most prevalent substances in our world is carbon. The carbon cycle supports the ecology of Earth. The exchange of carbon gases is crucial for sustaining a breathable environment. Human emissions (from production) and breathing release carbon. When an animal breathes, carbon dioxide waste products are released into the atmosphere due to respiration. Both human release and release from decaying plant and animal life contribute to respiration. Carbon builds up in the atmosphere and is reabsorbed by soil and plants (on land and in the water).

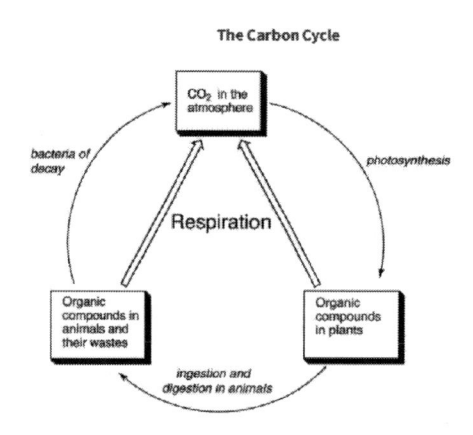

Meteorology:

> **LEARNING OBJECTIVES**
>
> In this section, you will learn to:
>
> identify layers of the Earth's atmosphere
>
> describe different types of fronts and clouds

Meteorology is the study of the atmosphere and atmospheric phenomena in general, in addition to the study of weather.

EARTH'S ATMOSPHERE:

First, you should be aware that the atmosphere is divided into multiple layers, starting right here at the Earth's surface, and going up several thousand kilometers above us. Layers include:

- **Troposphere:** The lowest layer is where all weather occurs. It is an area where air packets rise and fall. It can be anywhere from 6 to 17 kilometers thick, depending on the latitude and the time of year. The troposphere contains most of the Earth's atmosphere, composed primarily of nitrogen and oxygen (around 79 and 21 percent, respectively).

- **Stratosphere:** The stratosphere is located above the troposphere and is characterized by primarily horizontal wind. Ozone, a highly reactive form of oxygen, is present in high concentrations in the upper stratosphere's thin ozone layer. This layer absorbs most of the UV light from the Sun. Approximately 60°C is the temperature when you ascend into the stratosphere.
- **Mesosphere:** The mesosphere is located 90 kilometers above Earth and sits above the stratosphere. The temperature begins to plummet once more when you hit the mesosphere, reaching as low as 90°C. Here, we can observe "falling stars," meteoroids that strike the Earth and burn in the atmosphere.
- **Thermosphere:** Because there is little substance to reflect solar radiation above the mesosphere, temperatures rise with elevation. The thermosphere has experienced temperatures as high as 2,000°C.

FRONTS:

Wind and the movement of air masses with varying temperatures toward one another are caused by variations in air pressure. A warm front form when a warm air mass overtakes a cold one. The heavier cold air in front of the warm air passes over as it moves forward. Water vapor in the warm air rises and condenses to form clouds, creating rain, snow, sleet, or freezing rain—often all four—depending on the weather. A cold front form when a cold air mass overtakes a warm one.

Most cold fronts sweep across an area, followed by a line of precipitation. On the other hand, some cold fronts move with little or no precipitation. The rapid change in winds and temperature is the only indication that a front has passed through your area.

Occasionally, when two air masses collide, none is moved. Instead, there is a deadlock as the two fronts clash. We refer to this as a stagnant front. Cloudy, rainy weather frequently results from stationary fronts and might continue for a week or longer.

CLOUDS:

Clouds can be divided into several types based on size, shape, and height. There are three primary types:

- **Stratus:** Low-lying, wide, flat clouds called stratus cover the sky. Fog is the name for the lowest low clouds that touch the ground. Rain is imminent, according to the presence of dark stratus clouds.
- **Cumulus:** Cumulus clouds are large, puffy, popcorn-like clouds with rounded tops and comparatively flat bottoms. Expect heavy rain when cumulus clouds darken.
- **Cirrus:** Cirrus clouds are slender, wispy clouds that can be seen 20,000 feet or higher in the atmosphere.

Our Solar System:

> **LEARNING OBJECTIVES**
>
> In this section, you will learn to:
> - understand the structure of the solar system.
> - understand the relationship of the Earth to other planets and solar bodies.
> - understand how Earth's position relative to the Sun produces Earth's seasons.

One star, the Sun, eight planets and all their moons, thousands of asteroids (minor planets), and an equal number of comets make up our solar system.

THE SUN:

A G2V star, often known as a yellow dwarf, is what the Sun is. G2 stars are yellow, have surface temperatures of about 6,000 °C, and are rich in neutrally charged elements, including iron, magnesium, and calcium. The "V" denotes the Sun's status as a dwarf star or relatively tiny by stellar standards. The Sun is believed to be only slightly older than the Earth itself, with an estimated age of roughly 4.7 billion years.

Despite being a dwarf in relation to other stars, the Sun makes up over 99.9% of the solar system's mass. Like all stars, the Sun is a massive ball of plasma heated above its critical point by atomic processes. Our Sun's core is roughly 15,000,000 degrees Celsius, while the surface is between 4,000 and 15,000 degrees Celsius. The Sun has a surface area approximately 12,000 times larger than Earth and a diameter of about 1.4 million kilometers, or more than 100 times larger.

THE PLANETS AND OTHER PHENOMENA:

Mercury, Venus, Earth, and Mars are the four planets nearest the Sun. They are called terrestrial planets because they share many characteristics with Earth, including inner metal cores and rocky surfaces. The only terrestrial planets with moons are Earth and Mars; however, the moons of Mars are significantly smaller than the Moon. Therefore, the largest of the terrestrial planets is Earth.

The term "outer planets" refers to the four planets that lie beyond Mars: Jupiter, Saturn, Uranus, and Neptune. Additionally, they all have rings, with Saturn having the most. Most Saturnian rings and probably those of the other planets are made of ice crystals.

Scientists no longer classify Pluto, initially regarded as the ninth planet, as a real planet. However, a mnemonic (or memory technique) for the planets may have been taught to you in school; here is one that excludes Pluto: Just now, my highly educated mother served us nachos. (Pluto is shown in the following diagram but remember that it is no longer considered a planet. Also, keep in mind that the diagram below is not scaled.)

The Solar System:

The solar system also has many tiny bodies, including comets and asteroids. Smaller pieces of asteroids and cometary debris are referred to as meteoroids. When they enter the Earth's gravitational field, they become "falling stars," or meteors, when they burn up in the planet's mesosphere.

Meteorites are meteoroids that make it to the Earth's surface. In between Mars and Jupiter, there is an asteroid belt. Beyond the known planets is the Kuiper Belt, a much larger assemblage of asteroids and other relics from the solar system's creation. Occasionally, comets are referred to as "dirty snowballs" or "icy mudballs." They are made up of dust that, for some reason, wasn't incorporated into planets when the solar system was forming, along with ice (including frozen gases and water). Except when they are close to the Sun, comets are invisible.

Comets have very visible tails, up to several hundred million kilometers long, made of plasma and laced with rays and streamers brought on by interactions with the solar wind when they are close to the Sun and active. Our own Moon is the most significant body in the solar system, next to the Sun, regarding life on Earth. We have tides because of the gravitational attraction between the Moon and Earth.

Twice daily, high tides occur when the Moon is at its closest and furthest points from the affected body of water. It is believed that without the regular ebbing and flowing of the marine tides in coastal locations, life would not have emerged on land—a yearly orbit of the Earth around the Sun.

The northern pole is slanted toward the Sun, and the southern pole is tilted away from it for a portion of the year due to the Earth's slight tilt on its axis. In the northern hemisphere, that time of year is winter, while in the southern hemisphere, it is summer. The seasons are reversed for a portion of the year; it is winter in the northern hemisphere and summer in the southern hemisphere during that time.

The Earth is tilted, so neither the northern nor southern poles are looking toward the Sun during the transitional period between summer and winter. The seasons are spring and autumn. Study the diagram that appears after.

Earth's Seasons:

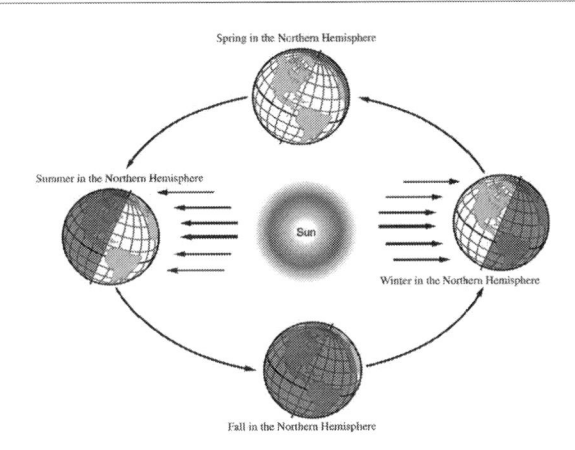

Imagine how frightened ancient people must have been when they could no longer see the Sun or the Moon during an eclipse. Since we know how eclipses happen, we can anticipate them well. When the Moon is directly between the Earth and the Sun during the day, a solar eclipse takes place.

The Sun may be entirely or partially hidden by the Moon's comparatively modest shadow on the surface of the Earth. However, solar eclipses are brief and travel quickly across the surface of the Earth because the Moon is moving through its orbit around the Earth, and the Earth is rotating on its axis.

There is a lunar eclipse when the Earth passes directly between the Sun and the Moon, casting a shadow on it. So long as the Moon is above the horizon, a lunar eclipse can be observed everywhere on Earth. Lunar eclipses can also be partial or total, just like solar eclipses. However, lunar eclipses linger much longer than solar because the Earth casts a wider shadow than the Moon.

Earth and Space Science Practice Questions:

1. The ozone layer is found in the following:
(A) troposphere
(B) stratosphere
(C) mesosphere
(D) thermosphere

2. What layer is located immediately beneath the Earth's crust?
(A) inner core
(B) plates
(C) mantle
(D) outer core

3. The clouds that occur at the highest altitude are called:
(A) cirrus
(B) cumulus
(C) nimbus
(D) stratus

4. The Kuiper belt is _____?
(A) a group of meteors that orbit around the earth
(B) a collection of asteroids and other objects left over from the formation of the solar system
(C) a layer of Earth's atmosphere
(D) a grouping of asteroids that orbit the sun between Mars and Jupiter

5. Which of the following is most responsible for the oceanic tides?
(A) the gravitational pull of the Sun on the Earth
(B) the gravitational pull of the Moon on the Earth
(C) the heat of the Sun
(D) the magnetic pull of the poles

6. Oxygen makes up approximately _____ percent of Earth's atmosphere.
(A) 10
(B) 21
(C) 78
(D) 90

7. The Richter scale is used to measure the intensity of _____.
(A) hurricanes
(B) blizzards
(C) earthquakes
(D) tornadoes

8. The sequence of the movement of water in the hydrologic cycle is:
(A) evaporation → transpiration → condensation → runoff/infiltration → precipitation
(B) runoff → condensation → infiltration → evaporation/transpiration → precipitation

(C) precipitation → evaporation/transpiration → condensation → runoff/infiltration
(D) precipitation → runoff/infiltration → evaporation/transpiration → condensation

9. Granite is an example of what type of rock?
(A) compound
(B) igneous
(C) metamorphic
(D) sedimentary

10. Earth is somewhat shielded from harmful ultraviolet radiation by the
(A) Van Allen belt
(B) thermosphere
(C) ozone layer
(D) troposphere

Answers and Explanations
EARTH AND SPACE SCIENCE PRACTICE QUESTIONS:

1. **B**
The ozone layer is found in the upper stratosphere.

2. **C**
The mantle lies beneath the Earth's crust.

3. **A**
Cirrus clouds are found at the highest altitude of all clouds.

4. **B**
A group of asteroids orbit the sun between Mars and Jupiter, but the Kuiper belt is located beyond the outermost known Planets.

5. **B**
The oceanic tides are caused by the gravitational pull of the Moon on the Earth.

6. **B**
Oxygen accounts for approximately 21% of Earth's atmosphere. Nitrogen makes up about 78% of the atmosphere.

7. **C**
Like the moment magnitude scale, the Richter scale is a logarithmic measure of the intensity of earthquakes. An increase of 1 unit on the Richter scale represents an increase by a factor of 10 in the intensity of an earthquake.

8. **D**
Precipitation from the sky falls to the ground, becoming surface runoff or infiltrating into the ground. Evaporation from bodies of water or transpiration from plants causes water vapor to rise into the sky, where it condenses to form clouds, and the cycle begins again.

9. **B**
Granite and other rocks formed by the hardening of molten magma are classified as igneous rocks.

10. **C**
The ozone layer, between the stratosphere and mesosphere, is primarily responsible for absorbing much of the ultraviolet radiation from the sun.

Part III
PHYSICAL SCIENCE:

Scientists would know very little about our solar system without understanding physics and chemistry, collectively known as the physical sciences. This section will cover those two disciplines as well as the system of measurement used in both.

Measurement:

Scientists use the metric system, not the British measurement units such as ounces, miles, and gallons. The metric system has been used throughout this chapter, but it's time for a more detailed examination.

The key idea of the metric system is to designate one base unit for every kind of measurement and then make bigger or smaller units by adding prefixes to it (groups of letters added to the beginnings of words, like *anti–* or *pro–*). For example, the base unit for length in metric is called the meter (m), which is just over a yard (about 39.4 inches).

A prefix is added to measure small things like firearm ammunition or large things like the distance between cities. For example, *Milli–* means 1/1000, so a millimeter (mm) 1/1000 is the size of a meter. A 9 mm handgun employs the metric system's common units and basic conventions related to the major temperature scales (Fahrenheit, Celsius, and Kelvin).

Beretta M9 (the US military's standard sidearm) fires rounds that are .009m in bullet diameter. A centimeter (cm) is 10 times as big as a millimeter, with about 2.54 cm in one inch. A kilometer (km) is 1,000 meters. Metric measurements typically have two or three-letter symbols (like mm, cm, and km), usually the prefix's first letter followed by the base unit's first letter.

Prefix	Symbol	Value Relative to Base Unit.
mega	M	10^5
kilo	k	10^3
hecto	h	10^2
deka	da	10^1
base(no prefix)	—-	1
deci	d	10^{-1}
centi	c	10^{-2}
milli	m	10^{-3}
micro	μ	10^{-6}

The most frequent prefixes are milli-, centi-, kilo-, and occasionally mega-. These are also the ones that are most likely to appear on the ASVAB. For completeness, the remaining prefixes on the table are included, but it is unnecessary to memorize them or any additional prefixes besides those mentioned here.

Knowing the prefix names and their symbols is a good idea because a calculation inquiry is more likely to use the symbol—in this case, km for kilometer—than the name.

When other measurements are made using the prefix system, the base unit for mass is grams (g). For example, the amount of specific minerals and nutrients per serving on a food package's nutritional information is typically listed in milligrams.

As a result, big objects can be measured in kilograms (kg), and extremely little things can be measured in milligrams (mg). One kilogram will weigh roughly 2.2 pounds (at the surface of the Earth), and there are approximately 28.3 grams in one ounce. Space in three dimensions is measured in volume.

A cubic centimeter (cc) or a milliliter (mL) are two terms that can be used to describe a cube (square box) that is one centimeter on a side. A liter is equal to 1,000 milliliters. Hence, the milliliter (M) definition automatically implies the liter (L): the prefix says a milliliter is one-thousandth of a liter.

A gallon contains roughly 3.79 liters, slightly more than a quart in liquid measure and equal to about 33.8 ounces. The unit of time is the second (s), although metric prefixes are used for lower quantities (such as milliseconds) as necessary. However, hours (h) and minutes (min) are also frequently used.

Lastly, the Celsius scale, often known as degrees centigrade, is the metric system's equivalent of temperature. Water freezes at 32 degrees Fahrenheit and boils at 212 degrees Fahrenheit, according to the Fahrenheit scale, which is more widely known in the United States. Water freezes at 0°C and boils at 100°C on the Celsius scale. The following general formulae can be used to convert values from the Fahrenheit scale to the Celsius scale or vice versa:

F=9/5 C + 32
C=5/9 F -32

Finally, there is one other temperature scale commonly used by scientists: the **Kelvin scale** or the **absolute zero scale**. Absolute zero is the temperature at which matter has no heat and its molecules are completely still; in theory, absolute zero is the lowest temperature possible. On the Kelvin scale, absolute zero is set at 0 K, equal to −273°C.

Otherwise, the Kelvin scale uses the same increments as degrees Celsius so that water freezes at 273 K and boils at 373 K. Note that there is no degree symbol when writing out temperatures in the Kelvin scale.

The unit types discussed here are not exhaustive. Different base units come up in other areas of science, but the beauty of metric standards is that the value of each prefix is a constant. Metric prefixes will be applied in new contexts elsewhere in this chapter.

Physics:

The study of the characteristics, transformations, and interactions of matter and energy is known as physics. The study of mechanics, thermodynamics, magnetism, optics, and electricity are only a few of the many subfields of physics. Therefore, only the physics that might be tested on the ASVAB's General Science portion will be covered in this review.

MASS *VS.* WEIGHT:

Words frequently used ambiguously in daily speech have very rigid definitions in physics. For instance, although they have very different definitions, mass and weight are frequently used synonymously.

Weight is the force that gravity applies to an object's mass, whereas mass is the quantity of substance that anything has. Despite being far from the nearest objects that produce gravity, a person can become almost weightless in outer space while still having all her bulk.

MOTION

Velocity is the rate at which an object changes position. Change in position is called **displacement,** and velocity is defined as the total displacement per unit time. Therefore, it can be calculated as **velocity = displacement of an object ÷ time**.

$$\vec{v} = \frac{\vec{d}}{t}$$

The term "vector quantity" in physics refers to a quantity that both a magnitude and a direction may fully define. For instance, a car heading west at 7.5 m/s (or miles per second) would be said to be moving at that speed if it covered fifteen meters in two seconds. These symbols frequently have a tiny arrow above them to indicate that displacement is also a vector. A vector is not time.

Momentum is a measure of the quantity of motion of an object. It corresponds to how difficult it is for a moving object to stop. The formula definition of momentum is **Momentum = mass × velocity**. In the symbolic version, momentum is represented by the letter *p* since *m* is already used for mass.

$$\vec{p} = m\vec{v}$$

This relationship means, for example, that a semitrailer truck moving at 5 km/h has more momentum than a person walking at the same speed and that you have more momentum when running than when walking.

Momentum is also a vector quantity, so two objects moving toward each other have opposite directions of momentum, which will partially or completely cancel out if they collide.

Acceleration is the rate of change of velocity. **Acceleration = change in velocity ÷ change in time**.

The Δ (delta) symbol represents change.

$$\vec{a} = \frac{\Delta \vec{v}}{t}$$

You can see an acceleration in a stopped vehicle when the light turns green, and the driver depresses the gas pedal. The movement of the speedometer needle shows acceleration as the car's velocity increases moment by moment until it plateaus at cruising speed. Acceleration is also a vector quantity.

FORCES AND ENERGY:

Force is the push or pull that causes an object to change its speed or direction.

Weight is just one example of a force; in this case, the force is due to gravity. A unit of force is called a **newton (N)**, which is the force required to impart an acceleration of one meter per second squared to a mass of one kilogram.

Work is performed on an object when an applied force is along the same line of movement. **Work = force × displacement**, where the directions of force and motion are parallel.

$$W = \vec{F}\vec{d}$$

A unit of work is called a **Newton-meter** or **joule (J)**. Performing work uses up **energy**, measured in joules, equal to the work performed. The reason we have to consume food regularly is that we are constantly using up energy: when we move, when our heart pumps blood, when our lungs inhale and exhale, when we generate warmth to maintain our body temperature, and so on. The nutritional information on food packages sometimes lists the food energy per serving in **kilojoules (kJ)** in addition to the traditional British unit of kilocalories (usually referred to, confusingly, as "calories" in everyday speech). **Power** is the rate at which work is performed or energy is converted. It's defined as the amount of work done or energy converted per unit of time and can be calculated as **Power = work ÷ time**, or **Power = (force × distance) ÷ time**.

P=w/t

The main power unit is the watt (W), where one watt is defined as one joule per second. Be sure not to mix up the symbol for the Watt unit with the symbol for work in the formula, nor the formula symbol for mass with the symbol for the meter unit. Units go with a number. The letter symbols in each formula stand for an unknown measurement value, which a number will replace during calculation.

NEWTON'S LAWS:

Isaac Newton, a famous physicist, formulated three laws of motion that describe the behavior of objects when they are in motion. These laws, commonly known as Newton's laws of motion, are fundamental to mechanics and essential for understanding various physical phenomena. This chapter will discuss these laws and their significance in physics.

Newton's First Law of Motion:
Also known as the law of inertia, Newton's first law of motion states that an object at rest will remain at rest, and an object in motion will continue to move at a constant velocity unless acted upon by an unbalanced force. In simpler terms, an object will maintain its state of activity until acted upon by force. The amount of inertia an object has depends on its mass. The greater the mass, the more difficult it is to change the object's motion.

Newton's Second Law of Motion:
Newton's second law of motion states that the acceleration of an object is directly proportional to the net force acting on it and inversely proportional to its mass. This law can be mathematically expressed as F = ma, where F is the net force, m is the object's mass, and a is the object's acceleration. In other words, the greater the force applied to an object, the greater its acceleration. The object's mass also plays a significant role in determining the acceleration. The larger the object's group, the smaller the acceleration for the same force applied.

Newton's Third Law of Motion:

Newton's third law of motion states that for every action, there is an equal and opposite reaction. This law implies that when two objects interact, they exert equal and opposite forces on each other. For example, if you push a wall with a force of 10 N, the wall will return with an equal and opposite force of 10 N. This law is essential for understanding the motion of objects in contact with each other, such as collisions.

The universal gravitational law of Newton. All objects in the cosmos are attracted to one another equally, and the strength of this attraction changes directly as a product of their masses and inversely as a square of their distance from one another. We refer to this force as gravity. The following equation, where G is a constant with a value of 6.67 1011, and r is the separation between the mass centers of the two objects, represents Newton's law of universal gravitation.

Take, for example, the gravitational force between the Sun and the Earth.

The following consequences follow from the law of universal gravitation:

- If the mass of the Earth were doubled, the force on the Earth would double.
- If the mass of the Sun were doubled, the force on the Earth would double.
- If the Earth were twice as far away from the Sun, the force on the Earth would be a factor of four smaller.
- The force exerted on the Earth by the Sun is equal and opposite to the force exerted on the Sun by the Earth (Newton's third law).

At the surface of the Earth, the acceleration due to gravity is 9.8 m/s2. This bears remembering and can be applied to the formula for Newton's second law to determine the weight (gravitational force) of any object given its mass.

ENERGY:

Energy can be characterized as the ability to perform work, as described in this chapter. The mechanical energy that can be kinetic or potential is the subject of several ASVAB energy questions. For example, the energy a moving item possesses is called kinetic energy. Potential energy is the energy that an object has stored in it due to its position, shape, or other characteristics.

Energy conservation law states that energy cannot be created or destroyed. Instead, it transforms between many forms. For instance, if a boulder is perched precisely on a cliff's edge, it has potential energy in relation to the ground below the cliff. The potential energy is transformed into kinetic energy immediately before the rock touches the ground if it is displaced and falls freely. Waves carry both sound and light energy, while light presents more challenges. So, let's examine the prop.

SOUND WAVES:

When an item vibrates, it disturbs the medium around it and sends ripples in all directions. This is how sound waves are made. These ripples, or waves, can move through air, liquids, and objects but not through a space vacuum. Sound waves that travel through water move faster than those that travel through the air, and sound waves that travel through water move faster than those that travel through metal or wood.

The frequency, or rate of shaking, of sound waves is directly linked to the pitch of the sound. When sound waves vibrate quickly, they have a high frequency. This makes the pitch of the sound high. Frequency is usually recorded in hertz (Hz), which is the number of times something happens every second. Humans can't hear sound waves with a very high pitch, which means many movements per second. Dogs and other animals can hear them, though.

A person's hearing range is usually between 20 Hz and 20 kHz. So, sometimes, a person or a dog will hear a sound at a different frequency than it is. The Doppler effect is to blame for this. The Doppler effect happens when either the source of the sound waves or the listener is moving closer together (making the pitch frequency sound higher than it is) or farther apart (making the pitch frequency sound lower than it is). The way the sound of a police or ambulance siren seems to change when it speeds by is a great example of the Doppler effect.

THE ELECTROMAGNETIC SPECTRUM:

It's important to remember that visible light is only a small part of the electromagnetic range (the part that we can see). All the different wavelengths and frequencies of energy are included in the electromagnetic spectrum. In the electromagnetic range, light waves are in the middle.

The electromagnetic spectrum goes from radio waves, which have the lowest frequency and longest duration, to microwaves, to infrared waves, to visible light, to ultraviolet light, to X-rays. Finally, gamma rays are the most energetic form of light.

Depending on the frequency of the waves, visible light also has different colors. Since red has the lowest frequency, wavelengths just below visible light are called infrared, and wavelengths just above visible light are called ultraviolet.

OPTICS:

As we've already said, light and electromagnetic range behave like waves. Light waves, on the other hand, are not made of matter like sound waves are. Instead, they are made of electricity and can move through space. They also move a lot faster than sound waves do. In a vacuum, the speed of light is 299,792,458 meters per second, about 186,000 miles per second or 300 million meters per second.

Refraction:

It's important to note, though, that the speed of light can change based on what it's going through. For example, light moves more slowly through water or glass than through a vacuum. The amount that light slows down in a certain medium is called its "refractive index." For example, a diamond has a refractive index of 2.4.

This means that light moves through a vacuum 2.4 times faster than it does through a diamond.

When light moves from one medium to another, the speed difference makes it turn, just like a car changes direction when it hits a slippery patch of road. Refraction is the name for this bending, and light bends at a bigger angle when the change in the index of refraction is larger.

Reflection:

Any wave, including light that hits a flat, smooth barrier follows the law of reflection, which says that the angle of impact is equal to the angle of reflection when measured from a line at a 90° angle to the barrier. This barrier is often a mirror when it comes to light.

Concave and Convex Mirrors and Lenses:

Mirrors and lenses can either be flat, or they can be concave (curved in, like a cave), or convex (curved out, like a swelling). Because of the law of reflection, a **concave mirror** is also called a **converging mirror** because the angles of incidence of rays of light parallel to the normal all converge upon a point.

Converging Mirror:

If you move the mirror away from the source of the image, the image gets blurry as the source gets closer to the mirror's focal point, where the angles of impact come together. As the source of the picture keeps moving past the focal point, the image keeps coming back in the mirror, but this time it is upside down.

On the other hand, a convex mirror is called a divergent mirror because it spreads out the light waves that hit it. Refraction, which is different from reflection, is how lenses work. A lens thicker in the middle than on the sides is called a convex lens. It is also called a converging lens because parallel waves that go through it come together. This kind of lens is used to fix farsightedness in reading glasses. It is also used in magnifying glasses, cameras, telescopes, and microscopes.

A concave lens is a thicker lens at the sides than in the middle. It is also called a diverging lens because the light waves that pass through it spread out. When a person is nearsighted, light waves come together before they reach their eye. As a result, a nearsighted person can only see things that are close to them. When put in front of the eye, a concave lens twists light to come together further back into the eye, where the retina is. This corrects nearsightedness.

HEAT:

There are three ways for heat to move from one thing to another: conduction, convection, and radiation. Heat always moves from places that are warmer to cooler places.

- **Conduction:** The easiest way for heat to move is through conduction. It's done by touching something directly, like putting your finger on a hot iron, which you shouldn't try at home. Most metals are good at transferring heat. On the other hand, Wood, Styrofoam, and plastic don't conduct heat well, making them good insulation.
- **Convection:** Heat is transferred by convection when hot particles in a stream move around. Convection is shown by the hot air rising from a campfire. Temperature differences cause convection, which is what makes ocean currents and winds happen.
- **Radiation:** When electric waves carry heat, this is called radiation. Radiation is the way that the Sun's heat moves through space.

MAGNETISM:

There is a north pole and a south pole on a simple magnet.

Like in a Hollywood love story, opposites do attract. If you try to bring two north poles of a magnet together or two south poles, they will push against each other, and you can feel that force. On the other hand, if you move one magnet's north pole toward another magnet's south pole, they will attract each other.

Because the Earth is magnetic and has a North Pole and a South Pole, a magnetic compass with a small, light magnet set on a nearly frictionless, nonmagnetic surface can be used to find north or south. On one end of the magnet, usually called a "needle," there is usually an arrow and the letter "N." This end of the needle is the south pole of the magnet. So, it always points toward the North Pole of the Earth, which lets the person reading the compass figure out where they are.

Chemistry:
ELEMENTS AND THE PERIODIC TABLE:

Chemistry is the branch of science that studies the structure and characteristics of matter and the chemical processes that produce or transform one type of matter into another. Anything with mass and space is matter, and depending on its molecular makeup, all forms of matter exhibit specific chemical characteristics.

Here are some definitions for the beginning:

- **Element:** A pure matter that cannot be divided into other types of matter via standard chemical processes. All known elements are listed on the Periodic Table of Elements (see the image on the following page), which is the building block of all matter.
- **Atom:** The smallest part of an element that nonetheless possesses the element's characteristics. Compounds can be created when one atom combines with comparable particles of other elements. Atoms comprise a complicated arrangement of moving electrons revolving around a positively charged nucleus that contains protons and neutrons (except hydrogen).
- **Proton:** a subatomic particle with a positive electric charge in an atom's nucleus.

- **Neutron:** A subatomic particle known as a neutron, which is neutral because it lacks an electric charge, is present in the nucleus of an atom.
- **Electron:** is a subatomic particle that revolves around an atom's nucleus. The mass of an electron is extremely small compared to that of the other subatomic particles (neutrons and protons are more than 1,800 times as hefty as an electron), and it carries a negative charge. An atom typically has the same number of negative electrons surrounding its nucleus as positive protons.
- **Molecule:** The smallest multi-atom particle of an element or compound that can exist in the free state and still retain the characteristics of the element or compound. The molecules of elements consist of two or more similar atoms; the molecules of compounds consist of two or more different atoms.

Through decades of research, scientists have organized all the known elements into the periodic table, conveying multiple information about each element.

Periodic Table of Elements:

You must comprehend a few concepts to read the Periodic Table of Elements. To begin with, elements are organized from left to right and from top to bottom to increase the atomic number. The number of protons in an element (and, since those two numbers are equal, the number of electrons) is represented by the element's atomic number. Periods refer to the different rows of items. The time represents the different orbits of electrons around atoms or shells.

To reduce the atom's energy, electrons typically occupy the lowest shell they can and only travel to higher shells once lower shells have been occupied (although this rule is occasionally broken). Each atom in the first period's top row has one electron shell. The electrons in every element in the second row (second period), and so on, have two shells. The maximum number of shells allowed at this moment is seven.

The names of the columns in the Periodic Table are also unique. A group is any of the components found in a column. The number of electrons in an element's outer shell is the same for all group members. An element's group dictates its chemical properties much more so than the time.

This is because the quantity of electrons required to finish an element's outer shell affects how it interacts with other elements to create molecules. For instance, the first column's Group I elements have one electron in their outer shell. Alkali metals, a term used to characterize soft, silvery metals that strongly react with water, comprise this group. The more aggressive this response is, the lower you go in the group.

Likewise, each element in Group II, the second column, has two electrons in its outer shell. Counting the columns, you may determine how many electrons are in the outer shell. Since their outer shells are filled, the noble gases—also referred to as inert gases—found in the far right-hand column often do not react with other elements.

Atomic Structure of Lithium and Sodium:

Li is in the second row of the periodic table because it has two electron shells, and Na is in the third row because it has three. However, both elements belong to Group I because they have one electron in their outer shell.

In addition to these crucial chemical characteristics, an element's atomic mass also affects where it appears on the Periodic Table. The atomic mass provided in each element's box indicates the average mass of a single atom. Why standard? Because the number of neutrons in some elements can vary, there may be slightly varied sizes of an element's atoms. Isotopes are the names for these various sizes.

Calculating the sum of the masses of the protons and neutrons yields a relatively accurate estimate of the atomic mass. The masses of each proton and neutron are very near to those of the so-called atomic mass unit (AMU). Since most hydrogen atoms only contain one proton, the atomic mass of hydrogen should be very near to one, which is at 1.007 amu. Look at the Periodic Table's box for a particular element right now. Cl2 is right here.

Chlorine:

The number 17 refers to the number of protons in the chlorine atoms nucleus and the number of electrons orbiting its shells. Unlike several other common elements, such as Fe (iron), Au (gold), Ag (silver), W (tungsten), or Na (sodium), which have atomic symbols derived from their Latin names rather than their well-known English names, Cl's symbol, Cl, is pretty like its name.

COMPOUNDS:

Unstable elements easily combine to generate compounds with characteristics quite different from the elements they are made of. For instance, the stable and edible chemical sodium chloride (NaCl), more widely known as table salt, is created when the exceedingly unstable and toxic elements sodium and chlorine mix. To produce a tightly bound crystalline structure when salt is solid, each chlorine atom takes an electron from each sodium atom, giving table salt its name as an ionic compound.

However, when dissolved in a liquid, like when table salt is added to water, the atoms separate into sodium and chloride ions. Each sodium atom is positively charged because it has lent an electron to its corresponding chloride atom, and each of the chloride atoms is negatively charged for the same reason. An ion is an electrically charged atom.

An illustration of a covalent compound is table sugar, which, among other things, does not ionize when dissolved in water. In a covalent compound, the molecules' atoms share electrons in pairs so that each one contributes half of the pair's energy, and both atoms strongly hold the pair. They won't separate as ions do as a result.

ACIDS AND BASES:

When dissolved in water, an acidic material releases positively charged hydrogen ions (H+). Although not all acids are safe to drink, they often have a sour flavor and erode metals. Vinegar, which contains acetic acid, and lemon juice, which contains citric acid, are two typical acidic solutions. Hydrochloric acid, nitric acid, and sulfuric acid (often used in lead batteries) are more potent acids that should all be handled carefully because they are highly corrosive.

When dissolved in water, a substance releases negatively charged hydroxyl ions (OH-), known as bases. Alkaline compounds are also referred to as basic substances. Bases usually have a bitter flavor.

Baking soda (sodium bicarbonate) and liquid soap (which frequently contains potassium hydroxide) are two examples of basic materials commonly found in a kitchen. However, bleach's most prevalent active ingredients, sodium hypochlorite, and sodium hydroxide, two more potent bases, are equally as caustic and hazardous as equivalent acids. When bases and acids interact (and they interact strongly), the two chemicals neutralize one another and transform into salt and water.

A solution's pH ranges from 0 to 14, representing how basic or acidic the solution is. With each lower number on the pH scale, acidity increases tenfold, classifying solutions with a pH lower than 7 as acidic. For instance, battery acid is 100 times more acidic than vinegar and has a pH of 1, black coffee has a pH of 5, vinegar has a pH of 3, and vinegar is 100 times more acidic than vinegar. A pH of 7 indicates a neutral solution. The pH of pure water is 7.

Common Substances on the pH Scale:

Solutions with a pH greater than 7 are considered basic, with each additional number doubling the amount of alkalinity. The pH of baking soda is just above 8. Borax, a common ingredient in cleaning products and detergents, has a pH of slightly over 9 and is about 10 times as basic as home bleach, which has a pH of over 12 and is 1,000 times more basic than borax and baking soda. A tenth of baking soda's alkalinity, or somewhat basic, human blood has a pH level just above 7.

PHYSICAL CHANGE:

A physical change or a chemical change can occur to matter. Physical changes to matter can affect its size, shape, and form, not its molecules. Physical modifications include creating a sandcastle or molding a piece of plastic into a different shape. However, the arrangement of matter has changed. The atoms and molecules that make up matter are still the same. In other words, its chemical composition has not changed.

Phase transitions are one of the most significant types of physical change. When matter changes from one state of matter, such as solid, liquid, or gas, it enters a phase transition.

Compared to the liquid and gaseous forms, the solid-state exists at lower temperatures (although the precise temperature ranges vary depending on the element or compound). Atoms or molecules are tightly packed and immobile when a substance is solid. Solids keep their volume and shape constant. Ice cubes, which are composed of water molecules in their solid state, are a good example.

The liquid form can exist over a wider temperature range than the solid-state equivalent of the same element or compound. In a liquid, molecules can freely move around one another. Although their volume remains constant, liquids do not keep their shape. If left uncontrolled, they will either assume the shape of their container or will spread out as a puddle. For example, a faucet releases water in a liquid state.

The gaseous state occurs at higher temperatures than the first two states of matter. This is because molecules stretch out as much as they can in the gaseous state and flow even more freely from one another.

Gases don't keep their volume or shape constantly. If the room is available, a small amount of gas will spread out to fill a big volume, or it can be tightly compressed into smaller places (oxygen tanks contain a significant amount of tightly compressed oxygen gas). The typical names for water in a gaseous state are steam or vapor.

Sometimes, a direct transition between a solid and a gas can occur without needing a liquid state transition beforehand. Below is a list of the precise words for each phase transition.

CHEMICAL CHANGE:

New molecules of matter distinct from the initial molecules of matter are created during a chemical reaction. Chemical reactions are changes that occur chemically in elements and compounds. A chemical reaction produces several types of molecules by rearranging atoms into new configurations. Reactants are the molecules and atoms that start the reaction, while products are the molecules and atoms that come out of the reaction.

Since each water molecule contains two hydrogen atoms and one oxygen atom, water is also known as H_2O. Many water molecules will disintegrate in one chemical reaction, and the atoms will rearrange into hydrogen (H_2) and oxygen (O_2) molecules. This is how pure hydrogen and oxygen gas are produced from the reactant, water. Rust on iron and the transformation of wood into charcoal, carbon dioxide, and steam in a fire are two common chemical reactions.

Another example of the dramatic impact of chemical change is the interaction between sodium metal and chlorine gas. Both substances are exceedingly dangerous to people in their pure forms. With any moisture, sodium metal reacts and produces heat and sparks, making it impossible to contact with bare hands. When not in use, it is kept in oil. A deadly chemical weapon in the First World War was chlorine gas. However, these two hazardous compounds can undergo a chemical reaction that yields regular table salt, which is present in every kitchen.

Physical Science Practice Questions:

1. What is 77°F in degrees Celsius?
(A) 25°C
(B) 32°C
(C) 37°C
(D) 5°C

2. Which of the following best explains the recoil action of a shooting gun?
(A) Newton's first law of motion
(B) Newton's second law of motion
(C) Newton's third law of motion
(D) Newton's law of universal gravitation

3. The weight of a pendulum clock transforms potential energy into usable mechanical work. Assuming no energy loss, approximately how much work will be done if a 200 g weight descends 0.5 m?
(A) 0.1 J
(B) 1 J
(C) 100 J
(D) 1000 J

4. Which waves on the electromagnetic spectrum have the highest frequency?
(A) microwaves
(B) X-rays
(C) visible light
(D) radio waves

5. Pure water has a pH of
(A) 1
(B) 7
(C) 0
(D) 14

6. During a baseball game, two players attempt to catch the same ball. The 50 kg Player A is traveling west at 12 m/s, while the 60 kg Player B is traveling east along the same line at 10 m/s. What will happen when they collide?
(A) Player A will continue forward at reduced velocity, knocking back Player B.
(B) Player B will continue forward at reduced velocity, knocking back Player A.
(C) Player B will continue forward without any decrease in velocity, knocking back Player A.
(D) Player A and Player B will both come to a complete stop

7. A truck was driven meters. At the beginning of the trip, the fuel tank contained 40,000 mL of fuel; at the end, it still had 15,000 mL of fuel. What was the average fuel consumption, measured in liters per kilometer, for the trip?
(A) 0.0125 L/km
(B) 0.125 L/km
(C) 8.0 L/km
(D) 80 L/km

8. The apparent sound of a train's horn changes from a higher to a lower pitch as the train approaches and then passes a stationary observer. This phenomenon is called.
(A) a Newtonian shift
(B) sound wave compression
(C) the Doppler effect
(D) an auditory aberration

9. If an object is viewed through a converging lens, when will the object appear to be upside down?
(A) always
(B) never
(C) when the observer is closer to the lens than the focal length
(D) when the observer is farther from the lens than the focal length

10. If a block of solidly frozen carbon dioxide ("dry ice") is left in a container open to the air, some frozen carbon dioxide will become carbon dioxide gas without becoming a liquid first. This phase change is called _____.
(A) sublimation
(B) evaporation
(C) vaporization
(D) gasification

Answers and Explanations
PHYSICAL SCIENCE PRACTICE QUESTIONS:

1. A
To convert the temperature from degrees Fahrenheit to degrees Celsius, first subtract 32, and then multiply by 5/9
77-32=45
45x 5/9 =25

2. C
Newton's third law of motion explains the recoil action of a shooting gun: for every action, there is an equal and opposite reaction.

3. B
Using Newton's second law and the fact that acceleration near the surface of the Earth is 9.8 m/s2, the force of gravity can be calculated.
The mass must first be converted to kg units: m = 200 g = 0.2 kg. Then the force is (0.2 kg) × (9.8 m/s2) = 1.96 N. This force acts over a distance of 0.5 m, so W = (1.96 N) × (0.5 m) = .98 J. Answer choice (B) is the closest Match.

4. B
Of the wave states listed, X-rays have the highest frequency in the electromagnetic spectrum (gamma rays have an even higher frequency).

5. B

Pure water has a pH of 7, making it a neutral solution.

6. D

Determine each player's momentum. Player A has a momentum of 50 kg × 12 m/s = 600 kg·m/s in a westward direction. Player B has a momentum of 60 kg × 10 m/s = 600 kg·m/s in an eastward direction. Since the two players' momentums are equal in magnitude but opposite in direction, the total momentum is zero (Ptotal = (+600) + (−600) = 0 kg·m/s). To put it another way, since neither player has a greater quantity of motion than the other, neither one can "win" in a collision. If neither can overcome the other, both coming to a stop is the only possible option.

7. B

First, convert the quantities given to the units of measurement of the answer. Since *kilo–* is the prefix for 1,000, the truck traveled $2 \times 10^5 / 10^3 = 2 \times 10^2 = 200 \times 10^2 = 200$ kilometers (km). The amount of fuel consumed was 40,000 − 15,000 = 25,000 mL. The prefix *milli–* means one thousandth, so that converts to 25 liters (L). Since the desired units are liters/kilometer, divide 25 L by 200 km to get 0.125 L/km.

8. C

This is a classic example of the Doppler effect caused by the relative motion of a sound source and receptor.

9. D

Light waves converge at a distance equal to the focal length of a lens. Beyond that distance, the waves from the top of the observed object are below those from the bottom, making the object appear to be inverted.

10. A

When a liquid changes to a gas, that process evaporates under natural conditions and vaporizes if heat is applied. The transformation directly from solid to gas is called sublimation.

Part IV
GENERAL SCIENCE PRACTICE QUESTIONS:
GENERAL SCIENCE PRACTICE SET 1:

1. Which of the following best describes the difference between eukaryotic and prokaryotic cells?
(A) Prokaryotic cells do not have nuclei.
(B) Prokaryotic cells are only found in plants.
(C) Eukaryotic cells are only found in plants.
(D) Eukaryotic cells do not have nuclei.

2. Which of the following is a terrestrial planet?
(A) Mars
(B) Jupiter
(C) Saturn
(D) Neptune

3. When a skydiver jumps from a plane, she eventually reaches a maximum falling velocity of −56 m/s. After opening her parachute, her falling velocity decreases to about −11 m/s. What effect does the skydiver's parachute have on her motion?
(A) The parachute does not change the skydiver's velocity.
(B) The parachute applies a negative acceleration on the skydiver.
(C) The parachute applies a positive acceleration on the skydiver.
(D) The parachute applies a negative acceleration first and then a positive acceleration on the skydiver.

4. A freight elevator is used to lift loads of 12,000 N at a vertical distance of 250 m. The elevator takes 10 s to make this trip. Assuming no energy loss, what is the average power expenditure of the elevator when working?
(A) 300,000 W
(B) 3,000,000 W
(C) 12,000,000 W
(D) 30,000,000 W

5. The air in the troposphere is made up primarily of:
(A) oxygen
(B) helium
(C) carbon dioxide
(D) nitrogen

6. The cell membrane that surrounds an animal cell is a
(A) non-permeable structure
(B) semi-permeable structure
(C) fully permeable structure
(D) cell wall

7. At which level of the atmosphere does all weather take place?
(A) troposphere
(B) stratosphere

(C) mesosphere
(D) thermosphere

8. Which of the following is NOT true about the element hydrogen (H)?
(A) It is the only element with no neutrons.
(B) It is the lightest element.
(C) It is a noble (inert) gas and rarely reacts.
(D) It sometimes behaves like an alkali metal.

9. Which of the following activities results in a chemical change?
(A) mixing salt and sugar
(B) drying wet clothes
(C) boiling a pot of water
(D) burning pieces of wood

10. Airflow in the stratosphere is primarily:
(A) horizontal
(B) vertical
(C) clockwise
(D) counterclockwise

11. Iron is a mineral that is vital to red blood cell development. Foods that are good sources of iron include _____.
(A) meats, beans, and whole grains
(B) nuts and green leafy vegetables
(C) bananas, sweet potatoes, nuts, and seeds
(D) milk, yogurt, cheese, and spinach

12. If a pencil is placed in a clear glass of water and viewed from the side, the pencil will appear to bend sharply at the top of the water.

This is due to a phenomenon known as _____.
(A) elusive
(B) reflection
(C) refraction
(D) rotation

13. Red blood cells are produced in _____.
(A) the liver
(B) kidneys
(C) bones
(D) the spleen

14. Which of the following is NOT an example of phase change?
(A) fuel burning in a combustion engine
(B) dry ice changing directly from a solid to a gas
(C) sleet melting on a sidewalk
(D) morning dew appearing on a lawn

15. HIV (human immunodeficiency virus) is transmitted _____.
(A) by touch
(B) through the air
(C) by bodily fluids
(D) by proximity to an infected person

16. The Earth experiences seasonal patterns of weather because _____.
(A) its distance from the Sun varies at different times of the year
(B) solar wind activity varies
(C) its orientation to the Sun is tilted on its axis
(D) the Moon's position relative to the Earth varies

17. Body functions such as heart rate and digestion that take place without having to think about them are regulated by the _____.
(A) somatic nervous system
(B) cerebrum
(C) neurons
(D) autonomic nervous system

18. Which factor is most important in determining how atoms of an element will chemically react with other atoms?
(A) the number of electrons in the outer shell
(B) the number of neutrons in the nucleus
(C) the chemical symbol of the element
(D) the atomic mass

19. Our Sun is classified as what type of star?
(A) yellow dwarf
(B) white dwarf
(C) yellow giant
(D) white giant

20. Shale and sandstone are types of _____ rocks.
(A) igneous
(B) sedimentary
(C) metamorphic
(D) compound

21. A DNA (deoxyribonucleic acid) molecule is shaped like a _____.
(A) sphere
(B) double helix
(C) oblate spheroid
(D) triple helix

22. A net force of 15 N is applied to an object that starts at rest and accelerates to a velocity of 12 m/s in 4 seconds. What is the mass of the object?
(A) 3 kg
(B) 5 kg
(C) 12 kg
(D) 15 kg

23. A lunar eclipse is caused by _____.
(A) an anomaly in the Moon's orbit
(B) solar flares
(C) the Moon passing between the Earth and the Sun
(D) the Earth passing between the Sun and the Moon

24. Earthquakes occur due to the movement of plates in the Earth's _____.
(A) asthenosphere

(B) lithosphere
(C) outer core
(D) lower mantle

25. Which of the following organisms has an exoskeleton?
(A) the human being
(B) spider
(C) amoeba
(D) eagle

26. Our lungs convey _____ to blood cells and receive _____ and _____ from blood cells.
(A) oxygen, carbon dioxide, water
(B) oxygen, carbon dioxide, nitrogen
(C) carbon dioxide, oxygen, water
(D) nitrogen, carbon dioxide, oxygen

27. A fireplace is sealed behind a clear glass screen to prevent sparks and ashes from getting out. Nevertheless, a person sitting in front of the fireplace will be warmed by the heat of the fire primarily because some heat is still transferred by _____.
(A) convection
(B) radiation
(C) conduction
(D) osmosis

28. Sodium (Na), which is in Group 1 of the Periodic Table of the Elements, and chlorine (Cl), which is in Group 17, will bond together to form sodium chloride (NaCl). How will this compound behave when dissolved in pure water?
(A) The molecules will group to form large crystals.
(B) The molecules will separate into Na and Cl atoms.
(C) The molecules will separate into negatively charged sodium and positively charged chlorine ions.
(D) The molecules will separate into positively charged sodium ions and negatively charged chlorine ions.

29. Meteorology is the study of _____.
(A) meteors and meteorites
(B) whether
(C) the atmosphere and atmospheric phenomena
(D) the solar system

30. Human eggs are produced in the female's _____.
(A) uterus
(B) fallopian tubes
(C) cervix
(D) ovaries

Answers and Explanations:

1. **A**
Prokaryotic cells, such as bacteria, do not have nuclei. Their genetic material is contained within the central part of the cell body. Eukaryotic cells contain nuclei where all genetic material is held. Eukaryotic cells are plant, animal, fungi, and protist cells.

2. **A**
Terrestrial planets are those closest in composition to the Earth. Not surprisingly, they are also the inner planets and those closest in proximity to the Earth. Terrestrial planets have an inner core of metal and rocky surfaces like the Earth's, though their atmospheric characteristics can differ vastly from Earth's. Jupiter, Saturn, and Neptune are outer planets.

3. **C**
The direction of the skydiver's velocity is negative for the entire period discussed in the question. Still, due to the release of the parachute, the later velocity is smaller in magnitude. Since the parachute acts against the direction of motion of the skydiver (as established by the fact that the magnitude of the velocity is decreased instead of increased), the acceleration must be positive, opposite in sign to the negative velocity. Alternatively, this question can be reasoned out as follows: the skydiver is falling and has a negative velocity. Therefore, down is negative; parachutes slow down skydivers by pulling them upward. Therefore, the parachute provides positive acceleration.

4. **A**
The applied force of the elevator must counteract the weight of the load to ensure a safe, uniform velocity. So, the work done can be calculated as $W = (12{,}000 \text{ N}) \times (250 \text{ m}) = 3{,}000{,}000$ J. Power is the energy expended per unit of time, so $P = (3{,}000{,}000 \text{ J}) \div (10 \text{ s}) = 300{,}000$ W.

5. **D**
The air in the troposphere is approximately 78% nitrogen and 21% oxygen. This is the layer of air closest to the Earth's surface.

6. **B**
Animal cells have semi-permeable membranes which allow for osmosis. Osmosis is essential for maintaining homeostasis within the cell and preventing it from shrinking or bursting.

7. A
The troposphere is the lowest level of the atmosphere, where all weather affecting Earth occurs. This is the level where clouds form and from where precipitation falls.

8. C
The question asks for a statement about hydrogen, which is not true. Both answer choices (A) and (B) are true statements. (The atomic number of 1 means there is one proton, and that one proton accounts for the atomic mass, which is also 1). The question can now be answered by identifying one of the remaining statements as true or false. Since hydrogen is not a noble gas (helium is the lightest noble gas, with atomic number two), (C) is the correct answer. Alternatively, (D) can be identified as the last true statement (hydrogen has a single electron in its outer shell, like alkali metals), which by process of elimination means that (C) is the correct Answer.

9. D
Mixing salt and sugar is a physical change. When combined, the salt and sugar retain their molecular makeup. Drying wet clothes is also a physical change. When clothes are dried, water molecules evaporate, becoming a gas. The clothes and water molecules remain chemically unchanged. When boiling a pot of water, once again, liquid water molecules move into the gas phase, but there is no change in the chemical composition of the molecules. Finally, burning wood is a chemical change because the molecules in the wood are converted into new molecules that are different than the original.

10. A
Air flow in the stratosphere is primarily horizontal, unlike airflow in the troposphere, which has a strong vertical component.

11. A
Nuts and green leafy vegetables are good sources of magnesium. Bananas, sweet potatoes, nuts, and seeds contain potassium. Milk, yogurt, cheese, and spinach provide calcium. Meats, beans, and whole grains are all sources of iron.

12. C
The speed of light is slower in water than in air, causing light to bend at the interface between air and water. While this may look like what could be called an optical illusion, exclusivity is not a phenomenon related to light. Reflection would be due to light bouncing off a surface, such as a mirror. Rotation is irrelevant to this question.

13. C
The marrow inside human bones produces red blood cells. The liver has many functions, most related to the digestive system. Kidneys, in addition to their role in purifying blood, produce the hormone that stimulates red blood cell production in the bone marrow. The spleen stores and purifies the blood.

14. A
Burning is a chemical change, not merely a change of physical state. Although most materials do not transition directly from a solid to a gas, dry ice does, and the phase change is called sublimation. Sleet melting is an example of solid liquefying (melting). Morning dew results from water vapor in the air condensing on the grass.

15. C
HIV can only be transmitted by an infected person's bodily fluids, such as blood, saliva, or semen.

GENERAL SCIENCE PRACTICE SET 2:

16. C
Because of the tilt of the Earth, during the summer, the Sun's rays are more directly overhead, and the days are longer; hence, summer is the warmest season. Because of the tilt, summer occurs at "opposite" times of the year in the northern and southern hemispheres. The distance between the Earth and the Sun does vary, but the effect on climate is much smaller than the tilt. The Earth is farthest from the Sun during the months that those living in the northern hemisphere call summer. Solar winds are electrically charged particles that, while they may be disruptive to electronic communications, have little effect on climate. The Moon's position affects tides, not climate.

17. D
The somatic nervous system controls voluntary actions and sends sensory information to the brain. The cerebrum is the portion of the brain that is considered the center of intelligence. Neurons are individual nerve cells.

General Science (GS) | 55

18. A
Chemical reactions typically involve either the transfer or sharing of electrons among atoms. Neutrons are in the nucleus, which usually remains unchanged in chemical reactions (although the nucleus can be changed in a nuclear reaction). The chemical symbol of an element has absolutely no bearing on the element's properties. The atomic mass is made up almost of protons and neutrons, so chemical reactions have little to do with atomic mass. **19. A**

Our Sun is technically classified as a G2V star, more commonly referred to as a yellow dwarf because it is small relative to many other stars.

20. B
The gradual deposit of sediments forms sedimentary rocks over a very long time. The sediments could be sand, forming sandstone, or clay that could eventually become shale. Igneous rocks originate as molten magma, and metamorphic rocks are formed from other rocks altered by temperature, pressure, or chemical processes. The compound is not a recognized type of rock.

21. B
The two biopolymer strands of DNA coil around each other to form a double helix structure, which, when the covalent bonds between the strands are included, resembles a twisted ladder that bears no resemblance to a sphere or oblate spheroid. A triple helix configuration would be impossible since there are only two main strands.

22. B
Use Newton's second law of motion, $F = ma$, to solve this problem. The object reached a velocity of 12 m/s in 4 seconds, so the acceleration was $\frac{12}{4} = 3$ m/sec2. $15 = m(3)$, so $m = \frac{15}{3} = 5$ kg.

23. D
When the Earth is precisely in line between the Sun and the Moon, the Earth's shadow is cast over the Moon, resulting in a lunar eclipse. Eclipses have nothing to do with an anomaly in the Moon's orbit or solar flares. When the Moon is between the Earth and the Sun, the Moon's shadow falls on a portion of the Earth, resulting in a solar eclipse.

24. B
The lithosphere refers to the crust and upper mantle, the two layers closest to the surface. In order towards the center of the Earth from the lithosphere are the lower mantle, asthenosphere, and outer Core.

25. B
The prefix *exo–* means outside, as in "external." Human beings and eagles have endoskeletons inside their bodies. An amoeba is a single-celled organism that does not have a skeleton. A spider's outer shell serves as its skeleton.

26. A
Blood carries the oxygen that cells need to function to the cells and picks up the carbon dioxide and water that they generate, so the lungs have to support that process by supplying oxygen to the blood and carrying away carbon dioxide and water.

27. B
Heat is transferred by radiation when electromagnetic waves transfer heat, just as we are warmed by the Sun. Convection transfers heat by the actual movement of warmed matter, often air. If the fireplace were not sealed behind glass, heat would likely be transferred to a room by convection. Conduction of heat occurs when two bodies of different temperatures directly touch. While the air touching the screen will warm up and spread a very slight amount of heat into the space in which the fireplace is located, that heat transfer would be very small compared to the transfer by radiation. Osmosis, the passage of a fluid through a membrane, is unrelated.

28. D
Elements on the left side of the Periodic Table, such as those in Group 1, have weak bonds with the electrons in their outer shells and will give them up relatively easily. Elements in Group 17 are on the right side of the table and have a strong attraction for electrons. Therefore, compounds of atoms from these two groups will form ionic rather than covalent bonds. When compounds with ionic bonds are dissolved in water, they typically separate into charged ions. Since the Na atoms have given up an electron, they will become positively charged ions. The Cl atoms will retain an extra electron and become negatively charged Ions.

29. C
Watching meteorologists on TV presenting weather forecasts, one could assume that meteorology deals exclusively with the weather. However, science is much more broadly based, encompassing the atmosphere. Although choice (A) might be tempting, meteorology is unrelated to meteors.

30. D

The Latin word for eggs is "ova," so it makes sense that eggs would be produced in ovaries. From the ovaries, eggs travel through the fallopian tubes where, if they are fertilized, they may implant in the uterus. The cervix is located at the bottom of the uterus.

REVIEW AND REFLECT:

As you look back over your work in this chapter, think about these questions:

- Which concepts are you more comfortable with, and which seem less familiar?
- As you worked on practice questions in this chapter, were you careful to make predictions whenever possible?
- If you missed practice questions that called for calculations, was it because you didn't understand or remember the concept being tested or because you made a misstep while doing the math?
- Did you pay careful attention to the units required by a problem?
- Which science topics should you review again before Test Day?

CHAPTER 6

Arithmetic Reasoning (AR)

Tackling the Real World of Word Problems:

Test-takers often waste a lot of time reading and rereading word problems as if the answer might reveal itself to them by some miracle; however, correctly solving math word problems requires you to perform a series of organized steps:

1. Read the problem thoroughly.
2. Figure out what the question is asking.
3. Dig out the relevant facts.
4. Set up one or more equations to arrive at a solution and solve the problem.
5. Review your answer.

I cover these steps in detail throughout this section.

Reading the Entire problem:

The first step in solving a word problem is reading the entire problem to discover what it's all about. Next, try forming a picture of the problem in your mind or—better yet—sketch the problem on your scratch paper. Finally, ask yourself whether you've ever seen a problem like this. If so, what's similar about it, and what did you do to solve it in the past?

Figuring Out What the Question is Asking:

The second and most important step in solving a word problem is to determine precisely what the question is asking. Sometimes the question is asked directly. At other times, identifying the question may be a little more difficult. For example, suppose you're asked the following question:

What's the volume of a cardboard box measuring 12 inches long by 14 inches wide by 10 inches tall?

(A) 52 cubic inches
(B) 88 cubic inches
(C) 120 cubic inches
(D) 1,680 cubic inches

The problem directly asks you to determine the volume of a cardboard box. Recall from your high school algebra and geometry classes that the volume of a rectangular container is length, width, height, or *V lwh*—so 12 14 10 1,680. Therefore, the correct answer is Choice (D).

Now look at the next example:

How many cubic inches of sand can a cardboard box measuring 12 inches long by 14 inches wide by 10 inches tall contain?

(A) 52 cubic inches
(B) 88 cubic inches
(C) 120 cubic inches
(D) 1,680 cubic inches

This is the same problem, but the question you need to answer isn't as directly stated. Therefore, you must use clues embedded in the problem to determine the question. Would figuring out the perimeter of the box help you with this question? Nope. Would figuring out the area of one side of the box help you? Nope—you're not painting the box; you're filling it. The question wants you to determine the volume of the container.

Clue words can be a big help when figuring out which question is being asked. Look for the following clue words:

- » **Addition:** Sum, total, in all, perimeter, increased by, combined, added.
- » **Division:** Share, distribute, ratio, quotient, average, per, out of, percent.
- » **Equals:** Is, was, are, were, amounts to.
- » **Multiplication:** Product, total, area, cubic, times, multiplied by, of.
- » **Subtraction:** Difference, how much more, exceed, less than, fewer than, decreased.

Digging for the Facts:

After figuring out what question you're answering, the next step is determining which data is necessary to solve the problem and which data is extra. Start by identifying all the information and variables in the problem and listing them on your scratch paper. Ensure you attach measurement units contained in the problem (miles, feet, inches, gallons, quarts, and so on). After you've made a list of the facts, try eliminating those irrelevant to the question.

Look at the following example:

To raise money for the school yearbook project, Tom sold 15 candy bars, Becky sold 12, Debbie sold 17, and Jane sold the most at 50. How many candy bars did Becky, Debbie, and Jane sell?

The list of facts may look something like this:

- Tom 15 bars
- Becky 12 bars
- Debbie 17 bars
- Jane 50 bars.
- Total Becky, Debbie, and Jane sold.

Because the question is the total number of candy bars sold by Becky, Debbie, and Jane, the number of bars Tom sold isn't relevant to the problem and can be scratched off the list. Just add the remaining bars from your list. The answer is 79.

Setting Up the Problem and Working Your Way to the Answer:

It would help if you decided how to solve the problem and used your math skills to arrive at a solution. For instance, a question may ask the following:

Joan just turned 37. For 12 years, she's dreamed of traveling to Key West to become a beach bum. To finance this dream, she needs to save a total of $15,000. How much does Joan need to save each year to become a beach bum by her 40th birthday?

In mathematical terms, write down what the question asks you to determine. Because the question asks how much money Joan needs to save per year to reach $15,000, you can say y(years Joan has to save) m (money she needs to save each year) = $15,000. (Assume she saves the same amount each year.) Or, to put it more mathematically, $ym \$ 15\ 000$

You don't know the value of m (yet)—that's the unknown you're asked to find. But you can find out the value of y—the number of years Joan must save. If she's 37 and wants to be a beach bum by age 40, she has 3 years to save. So now the formula looks like this:

To isolate the unknown on one side of the equation, you divide each side by 3, so $3m/=15,000/$

Therefore, your answer is $m=5,000$

Joan needs to save $5,000 annually for 3 years to reach her goal of $15,000 by the time she's 40. You may be tempted to include the 12 years Joan has dreamed of this trip in your formula. However, this number was put into the problem as a distraction. It has no bearing on solving the problem.

Drawing a diagram

Sometimes, working through a problem is easier when you can visualize it. You can use as much scratch paper as you need on the ASVAB, so you may want to draw diagrams to help work out some of the questions.

For example, if you encounter a question that asks you how much tile you need to floor a six-walled room that measures 15 feet on one side, 14 feet on another, 8 feet on another, and 6 feet on a fourth, it can be really helpful to draw a diagram like this:

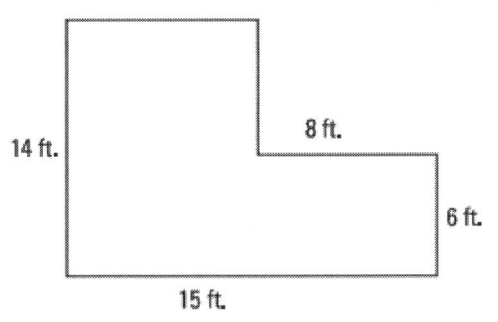

As you work through the practice questions in this book and complete the practice tests, use scratch paper to help you visualize what you're working through—that way, you'll know when drawing a quick sketch or creating a diagram will help you when you take the ASVAB.

Reviewing your answer:

Before marking your answer, sheet or punching that choice on the computer, review your answer to ensure it makes sense.

Review by asking yourself the following questions:

- » **Does your solution seem probable?** Use your common sense. If you determine that a 12-by-16-foot roof is covered in only 12 shingles, you've probably made a mistake in your calculations.
- » **Does it answer the question asked?** Reread the problem. For example, if a question asks you to calculate the number of trees *remaining* after 10% of the total was cut down, the correct answer wouldn't be 10% of the trees but rather the 90% still standing. Both answers will most likely be choices.
- » **Are you sure?** Double-check your answer. To keep you on your toes, test-makers often supply false answers that are very close to the correct answer.
- » **Is your answer expressed using the same units of measurement used in the problem?** A question may ask how many cubic feet of concrete are required to cover a driveway. Your answer in cubic yards would have to be converted to cubic feet so you can select the correct answer choice.

Although you may have been taught to round up for 5 or more and down for less than 5, rounding real-world problems requires a different mindset. For example, if someone needs 2.2 cans of paint for a particular job, they need 3 cans to ensure they have enough, even though you'd generally round down.

And if someone gets a 15-minute break for every 4 hours of work but works only 7 hours, they'd get only one break, even though 7 divided by 4 equals 1.75, generally rounded up to 2.

You may find that the solution you arrived at doesn't fit the facts presented in the problem. If this is the case, back up and repeat the steps until you arrive at a probable answer.

The Guessing Game: Putting Reason in Your Guessing Strategy

Guessing wrong on any ASVAB subtests doesn't count against you unless you guess incorrectly on a bunch of questions in a row at the end of the subtest when taking the CAT-ASVAB. If you don't guess, your chances of getting that answer right are zero, but if you take a shot at it, your chances increase to 25% or 1 in 4. Eliminate two wrong answers, and you have a 50-50 shot.

If you're taking the paper version of the ASVAB, you can always skip the tough questions and return to them after finishing the easier ones. However, if you're taking the computerized version of the ASVAB, the software won't let you skip questions, so you need to make your guess right then and there.

If you're taking the paper version of the test and elect to skip questions until later, make sure you mark the following answer in the correct space on the answer sheet. Otherwise, you may wind up wearing out the eraser on your pencil when you discover your error at the end of the test. Or even worse, you may not notice the mistake and get several answers wrong because you mismarked your answer sheet.

Using the Process of Elimination:

Guessing doesn't always mean "pick an answer, any answer." Instead, you can increase your chances of picking the correct answer by eliminating answers that can't be right.

To stop some wrong answers, you can do the following:

> » **Make sure the answer is realistic about the question asked.** For example, if a question asks you how much water would be required to fill a child's wading pool, 17,000 gallons isn't realistic. You can save time by eliminating this potential answer choice immediately.
> » **Pay attention to units of measurement.** For example, if a question asks how many feet of rope you'll need, answer options listed in inches or cubic feet are probably incorrect.
> » **Consider more accessible answer choices first.** Remember, you're not allowed to use a calculator on the ASVAB, so math answers you'd arrive at using complicated formulas are probably incorrect.

Solving What You Can and Guessing the Rest:

Sometimes, you may know how to solve part of a problem but not all of it. If you don't know how to do all the calculations—or don't have time for them—don't give up. You can still narrow down your choices by doing what you can.

Here's how partially solving problems can help:

> » When adding mixed numbers (a whole number and a fraction), add the whole-number parts first; then immediately eliminate answer choices too low. Or, when adding lengths, add total feet first and cross-off options that are too small, even before considering the inches.
> » Multiply just the last digits and cross off all answers that don't end in the correct numbers (assuming the answers aren't rounded).

Making Use of the Answer Choices:

If you're stuck on a particular problem, sometimes plugging possible solutions into an equation can help you find the correct answer. Using the answer choices can improve your guessing:

> » **Plug in each remaining answer choice until you get the correct answer.** Plugging in all the answer choices is time-consuming, so eliminate wrong decisions first.
> » **Estimate and plug in numbers that involve easy mental calculations.** For instance, if Choice (A) is 9 and Choice (B) is 12, plug in 10 and solve the equation in your head. Consider whether the right answer must be higher or lower than 10 and choose from there.
> » **Using a little logic, do calculations with a wrong answer choice.** Sometimes, a wrong answer choice—especially one that differs drastically from the other answers—represents an intermediate step in the calculations, so you can use it to solve the problem.

For instance, take this example:

A security guard walks six city blocks when they make a circuit around the building. If they walk at a pace of eight city blocks every 30 minutes, how long will it take them to complete a course around the building, assuming they don't run into thieves?

(A) 20.00 minutes
(B) 3.75 minutes
(C) 22.50 minutes
(D) 24.00 minutes

Choice (B) is way too low to be the correct answer, but it would be a logical guess for the security guard's rate for a single block. Multiply 3.75 minutes/block by 6 blocks, and you probably have a good candidate for the correct answer—22.50 minutes, Choice (C).

Arithmetic Reasoning (Math Word Problems) Practice Questions

Arithmetic Reasoning questions are math problems expressed in a story format. Your goal is to determine what the question is asking by picking out the relevant factors needed to solve the problem, setting up mathematical equations as needed, and arriving at the correct solution. Sounds easy. Try your hand at the following questions.

Easy Arithmetic Reasoning questions:

1. If a car travels 60 miles in 1.5 hours, what is its average speed in miles per hour?
A. 30 mph
B. 40 mph
C. 50 mph
D. 60 mph

2. A recipe calls for 3 cups of flour to make 24 cookies. How many cups of flour are needed to make 40 cookies?
A. 4 cups
B. 5 cups
C. 6 cups
D. 7 cups

3. A store offers a 15% discount on a jacket priced at $120. How much does the jacket cost after the discount?
A. $102
B. $108
C. $112
D. $118

4. If 5 workers can complete a task in 8 hours, how long would it take 10 workers to complete the same task?
A. 4 hours
B. 6 hours
C. 8 hours
D. 16 hours

5. A rectangular room measures 20 feet in length and 15 feet in width. What is the area of the room in square feet?
A. 200 sq ft
B. 300 sq ft
C. 350 sq ft
D. 400 sq ft

6. If the price of a shirt is $20 and the sales tax is 5%, what is the total price of the shirt including tax?
A. $21
B. $22
C. $23
D. $24

7. A train travels 300 miles in 5 hours. What is the average speed of the train in miles per hour?
A. 50 mph
B. 60 mph
C. 70 mph
D. 80 mph

8. A bookshelf has 4 shelves, with 15 books on each shelf. How many books are on the bookshelf in total?
A. 40
B. 50
C. 60
D. 60

9. If a rectangle's length is twice its width and the perimeter is 60 feet, what are the dimensions of the rectangle?
A. Length 20 ft, Width 10 ft
B. Length 15 ft, Width 30 ft
C. Length 10 ft, Width 20 ft
D. Length 30 ft, Width 15 ft

10. A group of 4 friends splits the cost of a $200 meal equally. How much does each person pay?
A. $40
B. $50
C. $60
D. $80

11. A bus travels 240 miles in 4 hours. What is its average speed?
A. 50 mph
B. 60 mph
C. 70 mph
D. 80 mph

12. If a tank holds 500 gallons of water and is currently filled to 80% of its capacity, how many more gallons are needed to fill the tank?
A. 100 gallons
B. 200 gallons
C. 300 gallons
D. 400 gallons

13. If 8 oranges cost $4, how much do 20 oranges cost?
A. $5
B. $8
C. $10
D. $12

14. A rectangle has an area of 48 square feet and a length of 8 feet. What is its width?
A. 4 feet
B. 6 feet
C. 8 feet
D. 10 feet

15. A worker earns $15 per hour and works 40 hours a week. What is his weekly salary?
A. $500
B. $600
C. $700
D. $800

16. What is the median of the following set of numbers: 10, 15, 20, 25, 30?
A. 15
B. 20
C. 25
D. 30

17. If a rectangle's area is 56 square feet and its length is 8 feet, what is its width?
A. 4 feet
B. 5 feet
C. 6 feet
D. 7 feet

18. A bag contains 4 red balls and 6 blue balls. What is the probability of picking a red ball at random?
A. 1/5
B. 2/5
C. 3/5
D. 4/5

19. If 3 t-shirts cost $45 in total, how much would 5 t-shirts cost?
A. $65
B. $75
C. $85
D. $95

20. A car travels 150 miles using 5 gallons of gas. What is the car's fuel efficiency in miles per gallon?
A. 25 mpg
B. 30 mpg
C. 35 mpg
D. 40 mpg

21. A square garden has a perimeter of 48 feet. What is the length of each side of the garden?
A. 10 feet
B. 12 feet
C. 14 feet
D. 16 feet

22. If a bicycle wheel rotates 500 times to travel 1 mile, how many times does it rotate in a 3-mile journey?
A. 1,000
B. 1,500
C. 2,000
D. 2,500

23. A store sells a notebook for $1.50 and a pen for $0.75. How much would 3 notebooks and 2 pens cost in total?
A. $3.50
B. $4.50
C. $5.50
D. $6.50

24. If 8 workers can build a wall in 10 hours, how long will it take 4 workers to build the same wall?
A. 5 hours
B. 10 hours
C. 15 hours
D. 20 hours

25. A box contains 12 pairs of socks. How many individual socks are there in the box?
A. 12
B. 24
C. 36
D. 48

26. If a triangle has a base of 10 feet and a height of 6 feet, what is its area?
A. 20 sq ft
B. 30 sq ft
C. 40 sq ft
D. 60 sq ft

27. A train ticket costs $45. If a group of 4 people buy tickets, how much will they pay in total?
A. $160
B. $180
C. $200
D. $220

28. If you divide a number by 4 and subtract 3, the result is 5. What is the number?
A. 20
B. 24
C. 28
D. 32

29. A pie is cut into 8 equal slices. If a family eats 3 slices, what fraction of the pie is left?
A. 1/8
B. 3/8
C. 5/8
D. 7/8

30. If you buy 2 shirts for $25 each and get a third shirt for half price, what is the total cost?
A. $50
B. $62.50
C. $75
D. $87.50

31. A store sells apples at $1.20 per pound. How much would 5 pounds of apples cost?
A. $4.00
B. $5.00
C. $6.00
D. $8.00

32. A rectangular field is 100 feet long and 50 feet wide. What is the perimeter of the field?
A. 150 feet
B. 200 feet
C. 250 feet
D. 300 feet

33. If a car's odometer reads 25,250 miles and the driver travels another 475 miles, what will the odometer read?
A. 25,625 miles
B. 25,725 miles
C. 26,000 miles
D. 26,225 miles

34. A clock shows 3:15 PM. What angle is formed by the hour and minute hands?
A. 0 degrees
B. 7.5 degrees
C. 15 degrees
D. 22.5 degrees

35. If a factory produces 480 widgets in 8 hours, how many widgets does it produce per hour?
A. 40
B. 50
C. 60
D. 70

36. A book has 300 pages and you read 5 pages each day. How many days will it take to read the book?
A. 50 days
B. 60 days
C. 75 days
D. 100 days

37. A circle has a radius of 7 inches. What is the area of the circle? **(Use π≈3.14)**
A. 43.96 sq in
B. 153.86 sq in
C. 154 sq in
D. 200 sq in

38. If 9 gallons of paint cover 360 square feet, how much area can 1 gallon of paint cover?
A. 30 sq ft
B. 40 sq ft
C. 50 sq ft
D. 60 sq ft

39. A jar contains 120 marbles, some red and the rest blue. If there are 72 blue marbles, how many are red?
A. 28
B. 38
C. 48
D. 58

40. If a cube has an edge length of 4 inches, what is the volume of the cube?
A. 48 cubic inches
B. 60 cubic inches
C. 64 cubic inches
D. 72 cubic inches

Answer Key with Explanations

1. B. Speed = Distance ÷ Time. Thus, 60 miles ÷ 1.5 hours = 40 mph.

2. B. Use the proportion: 3 cups/24 cookies = x cups/40 cookies. Solving for x gives x = 5 cups.

3. A. Discount = 15% of $120 = $18. Discounted price = $120 - $18 = $102.

4. A. If 5 workers take 8 hours, 10 workers (twice as many) would take half the time, so 4 hours.

5. B. Area = Length × Width. Thus, 20 ft × 15 ft = 300 sq ft.

6. A. Sales tax = 5% of $20 = $1. Total price = $20 + $1 = $21.

7. B. Speed = Distance ÷ Time. Thus, 300 miles ÷ 5 hours = 60 mph.

8. C. Total books = Number of shelves × Books per shelf = 4 × 15 = 60.

9. A. Let width = x. Then length = 2x. Perimeter = 2(length + width) = 60, so 2(2x + x) = 60. Solving gives x = 10 ft (width) and 2x = 20 ft (length).

10. B. Each person pays $200 ÷ 4 = $50.

11. B. Speed = Distance ÷ Time. Thus, 240 miles ÷ 4 hours = 60 mph.

12. A. 80% of 500 gallons = 400 gallons. To fill, need 500 - 400 = 100 gallons more.

13. C. If 8 oranges cost $4, then each orange costs $0.50. Thus, 20 oranges cost 20 × $0.50 = $10.

14. A. Area = Length × Width. So, 48 sq ft = 8 ft × width. Width = 48 ÷ 8 = 6 ft.

15. B. Weekly salary = Hourly rate × Hours worked = $15/hour × 40 hours = $600.

16. B. The median of a sorted list of numbers is the middle number. Here, the middle number is 20.

17. D. Area = Length × Width. So, 56 sq ft = 8 ft × Width. Width = 56 ÷ 8 = 7 feet.

18. B. Probability = Number of favorable outcomes ÷ Total number of outcomes. Here, 4 red balls out of 10 total balls = 4/10 = 2/5.

19. B. Cost per t-shirt = $45 ÷ 3 = $15. For 5 t-shirts, cost = 5 × $15 = $75.

20. B. Fuel efficiency = Distance ÷ Fuel used. So, 150 miles ÷ 5 gallons = 30 mpg.

21. B. Perimeter of a square = 4 × side length. Thus, 48 feet = 4 × side length, so each side is 12 feet.

22. B. Rotations in 3 miles = 3 × rotations in 1 mile = 3 × 500 = 1,500.

23. B. Cost = 3 notebooks × $1.50 + 2 pens × $0.75 = $4.50 + $1.50 = $6.00.

24. D. If 8 workers take 10 hours, 4 workers (half as many) would take double the time, so 20 hours.

25. B. There are 12 pairs, and each pair has 2 socks. So, total socks = 12 pairs × 2 = 24 socks.

26. B. Area of a triangle = 1/2 × base × height. So, 1/2 × 10 ft × 6 ft = 30 sq ft.

27. B. Total cost = 4 tickets × $45 = $180.

28. D. Let the number be x. According to the problem, x/4 - 3 = 5. Solving for x gives x = 32.

29. C. If 3 slices are eaten, 8 - 3 = 5 slices are left. Fraction left = 5/8.

30. B. Total cost = 2 shirts at $25 each + 1 shirt at half price ($12.50) = $50 + $12.50 = $62.50.

31. C. Total cost = Price per pound × Number of pounds. So, $1.20 × 5 = $6.00.

32. D. Perimeter of a rectangle = 2 × (Length + Width). So, 2 × (100 ft + 50 ft) = 300 feet.

33. B. New odometer reading = Initial reading + Additional miles. So, 25250 + 475 = 25725 miles.

34. B. At 3:15, the hour hand is a quarter way between 3 and 4. The angle between each hour mark is 30 degrees. Thus, the angle is 30° × 1/4 = 7.5 degrees.

35. C. Widgets per hour = Total widgets ÷ Hours. So, 480 ÷ 8 = 60 widgets per hour.

36. B. Days to read = Total pages ÷ Pages per day. So, 300 ÷ 5 = 60 days.

37. B. Area of a circle = π × radius^2. So, 3.14 × 7^2 ≈ 153.86 sq in.

38. B. Area covered per gallon = Total area ÷ Gallons of paint. So, 360 sq ft ÷ 9 gallons = 40 sq ft/gallon.

39. C. Red marbles = Total marbles - Blue marbles. So, 120 - 72 = 48 red marbles.

40. C. Volume of a cube = Side length^3. So, 4^3 = 4 × 4 × 4 = 64 cubic inches.

Chapter 7

Word Knowledge (WK)

For basic fighting training, you need a good vocabulary that helps you do well on the ASVAB's Word Knowledge subtest. Not only do you have to be able to spell (so you can tell words apart), but you also must know what the words on the test mean. The military wants to know how large your vocabulary is to determine if you can talk to other troops, understand written and spoken orders, report to higher-ups, and lead younger servicemembers.

But what if you don't know what a gun is and what carbon is? Don't worry too much; I'm here to help (or flood you with phrases to help you improve your vocabulary if that's what you want). With the help of this chapter and a little work, you can get a good handle on words. Then, you can check out the practice questions at the end of the chapter to see how well you've learned.

On the computerized form of the ASVAB, you have 9 minutes to answer 15 questions on the Word Knowledge subtest. This gives you 36 seconds to answer each question. (If your Word Knowledge subtest has practice questions, I explain in Chapter 1, you have 18 minutes to answer 30 questions.) So, the plan is a little tighter on the paper copy.

You have 11 minutes to answer 35 questions, meaning you have less than 20 seconds for each. Either way, you have plenty of time, as long as you stay focused and don't waste time thinking about your future in the military (I mean getting lost in thought as you think about joining the military).

Most candidates take the computerized form of the ASVAB these days. This version is made to test your skills quickly by asking you questions. You start with an about-average question.

If you give the right answer, you get a harder question. If you give the wrong answer, you get an easy question. Some people find that the ASVAB is easier or harder than they thought it would be. This is sometimes because the test changes to fit the person taking it.

Understanding How Important It Is to Know Words:

The ASVAB tests word knowledge because words stand for ideas, and the more words you understand, the better you can talk to others. The military wants to ensure that you can follow instructions for using and taking care of tools, rules and regulations written in military manuals, and what your superiors tell you. They also want to ensure you can talk to the troops who work under you.

You need a good vocabulary to get more supplies (by submitting necessary logistical requisitions) or get the job you want (by applying for personnel career-enhancement programs). Clear communication is so important to the military that it is taught and tested at every level of training, even basic training.

The Word Knowledge subtest is one of the four most important parts of the ASVAB. The other three are Paragraph Comprehension, Math Knowledge, and Arithmetic Reasoning. This subtest makes up a big part of your AFQT score, deciding whether you can join the military. It would help if you did well on the Word Knowledge subtest for many military jobs, like air traffic director, military intelligence, and even firefighting.

Military Line Scores That Use The Word Knowledge Score

Branch of service	Line Score
U.S. Army	Clerical, combat, general technical, operators and food, skilled technical, and surveillance and communication
U.S. Air Force	Administrative, General, and Mechanical
U.S Navy/Coast Guard	Administrative, General Technician, Hospital, Nuclear, and Operations
U.S. Marine Corps	Clerical and General Technician

Trying Out the Word Knowledge Question Format:

The Word Knowledge portion of the ASVAB measures your vocabulary. The questions usually come in one of two types:

> » The first type asks for a straight definition.
> » The second type gives you an underlined word used in the context of a sentence.

When you're asked for a straight definition, your task is quite simple: Choose the answer closest in meaning to the underlined word. Look at the following example:

- Abate most nearly means:
  (A) encourage.
  (B) relax.
  (C) obstruct.
  (D) terminate.

Abate means to suppress or terminate. In this case, the correct answer is Choice (D).

When you see an underlined word in a sentence, your goal is to choose the answer closest in meaning to the underlined word. **Remember:** *Closest in meaning* doesn't mean *the same thing*.

You're looking for similar or related words. For example:

- The house was derelict.
  (A) solid.
  (B) run-down.
  (C) clean.
  (D) inexpensive.

Here, the answer is Choice (B).

Tip:

> Remember that although you may know the word in the question, you may not know one or more words in the multiple-choice answers. If this is the case, use the process of elimination to help you narrow down your choices. Eliminate the words you know are wrong, and then choose the one most likely correct from what's left.

How to Build Words from Scratch:
Tips to Help You Figure Out What Words Mean:

There are more than 470,000 words in Webster's Third New International Dictionary. You don't have to learn all of them, which is good news. Even if you've never heard a specific English word, you can figure out what it means.

It takes time—often years—to build up considerable knowledge. Even though you don't have much time to study, that doesn't mean you should give up. Instead, pay attention to the tips in this part to help you get a better score on Word Knowledge.

From the Start to the End, Knowing Prefixes and Endings are Important:

Words comprise three main parts: prefixes, stems, and suffixes. Not every word has all three, but most have at least one. The meaning of a word is changed by its prefix, which is the first letter. The last letter of a word is called a suffix, and it also changes what the word means.

Roots are the parts of a word that are in the middle. Think of roots as the base of the word and prefixes and suffixes as parts of the word that are added to the base to change its meaning. (The next section has more information about, you got it, roots.)

These basic word parts generally have the same meaning in whatever word they use. For instance, the prefix pro- means in favor of, forward, or positive, whether you use it in the word proton or proceed. These tables list some common prefixes and suffixes. Each list has the word part, its meaning, and one word that uses each word part. Write down additional words that use each word part to help you memorize the list.

PREFIXES:

Prefix	Meaning	Sample Word
a-	no, not	atheist
ab- or abs-	away, from	absent
anti-	against	antibody
bi-	two	bilateral
circum-	around	circumnavigate
com- or con-	with, together	conform
contra- or counter-	against	contradict
de-	away from	detour
deca-	ten	decade
extra-	outside, beyond	extracurricular, extraordinary
fore-	in front of	foretell
geo-	earth	geology
hyper-	above, over	hyperactive
il-	not	illogical
mal- or male-	evil, bad	malediction
multi-	many	multiply
ob-	blocking	obscure
omni-	all	omnibus
out-	external	outside
que-, quer-, or ques-	ask	question, query
re-	back, again	return
semi-	half	semisweet
super-	Over, more	superior
tele-	far	telephone
trans-	across	transatlantic
un-	not	uninformed

Suffixes:

Suffix	Parts of speech it indicates	Meaning	Original Word: Suffixed Word
-able or -ible	Adjective	capable of	agree: agreeable
-age	Noun	act of	break: breakage
-al	Adjective	relating to	function: functional
-ance or -ence	Noun	an instance of an action	perform: performance
-ation	Noun	action, process	liberate: liberation
-en	Adjective, verb	made from	silk: silken
-ful	Adjective	full of	help: helpful
-ic	Adjective	relating to, like	anemia: anemic
-ical	Adjective	possessing a quality of	magic: magical
-ion	Noun	result of, act of	legislate: legislation
-ish	Adjective noun	resembling	child: childish
-ism	noun	condition or manner	hero: heroism
-ist	noun	one who or that which does	anarchy: anarchist
-ity	noun	quality of or state of	dense: density
-less	Adjective noun	not having	worth: worthless
-let	noun	small one	book: booklet
-ly	Adjective, Adverb	like	fearless: fearlessly
-ment	noun	act or process of	establish: establishment
-ness	Adjective, noun	possessing a quality	good: goodness
-or, -er	noun	one who does a thing	orate: orator
-ous	Adjective	state of	danger: dangerous

Suffixes are small parts of words that do a lot. They change the meaning of a word and generally tell you what part of speech a word is. This is helpful when you can't answer a Word Knowledge question immediately.

If you learn prefixes, endings, and roots, it will be easier to figure out the meaning of a word on the ASVAB that you don't know. People with big languages make them even bigger by figuring out what words they don't know to mean. (They also look things up in a dictionary.)

Getting to the Bottom of a Problem:

Root words are the parts of a word that make it up. With a slight change, one English root word can be a name, a verb, an adjective, or an adverb. If you know where a word comes from, you can usually figure out what it means, even if you don't know it.

For example, you can figure out what the word connection means if you know the root word attach. Likewise, you can determine adherence's meaning if you know what follow means.

Here are listed some common roots. Memorize them.

You'll be glad you did when you sit down to take the ASVAB.

ROOTS:

Root	Meaning	Sample Word
anthro or anthrop	Relating to humans	anthropology
bibl	Relating to books	bibliography
brev	short	abbreviate
cede or ceed	go, yield	recede
chrom	color	monochrome
cogn	know	cognizant
corp	body	corporate
dict	speak	diction
domin	rule	dominate
flu or fluc	flow	influx
form	shape	formulate
fract or frag	break	fragment
graph	writing	biography
junct	join	juncture
liber	free	liberate
lum or lumen	light	illuminate
oper	work	cooperate
path or pathy	Suffer, feeling	pathology
port	carry	portable
press	squeez	repress
scrib or script	write	describe
sens or sent	feel	sentient
tract	pull	traction
voc or vok	call	revoke

Word Families: Finding Similar Things:

When you see an unfamiliar word in the Word Knowledge section, don't get upset and pound on the computer (they make you pay for those things if you break them). After all, you may know the word—just in a different form. Suppose you run across the word *beneficent* on the Word.

Knowledge portion:

Beneficent most nearly means:
(A) kind.
(B) beautiful.
(C) unhappy.
(D) troubled.

You start to sweat as you sit there. If you've never seen the word before, you're done, right? Not necessarily. Pay more attention. What other word can you think of that starts with benefit? What about the word advantage? Something that helps is gain. The word benevolent likely has something to do with helping or helping out. So, when you look at your options, pick the one that has something to do with helping.

But wait! Help or aid are not mentioned in the answers. What's next? Just use the removal method. If someone is kind and helpful, they are usually not upset or unhappy. They might be pretty, but it's more likely that they're nice. So, Choice (A) is the best answer.

Remember when the guidance director at your high school told you to take French or Spanish? You should thank them if you do well on this part of the test. Why? Knowing other languages can help you figure out what English words mean.

For example, if you know that "health" in Spanish is "salud," you might be able to figure out what "salutary" means in English, which means "good for or promoting health." If you know that sang means "blood" in French, you might be able to figure out what sanguine means in English. Try to figure it out independently and then look it up in a book to see how close you got.

Taking Words Apart:

Pulling apart words is a great strategy when unsure what something means. However, even knowing what part of a word means can help you make a smarter choice—and on the ASVAB, every question counts.

Try to pull apart the word in this question to see whether you can figure out its meaning.

Detractor most nearly means:
(A) critic.
(B) driver.
(C) expert.
(D) adulatory.

Take apart the word *detractor:*

> » **de- is the prefix**
> » **tract is the root**
> » **-or is the suffix**

If you've learned what any part of the word means, whether it's the prefix, root, or suffix, finding the correct answer is easier

How to Use Synonyms and Antonyms: "Yin and Yang"

A synonym is a word that means the same or very close to what another word means.

Smile and grin mean the same thing. They are very close even though they don't mean precisely the same thing. An antonym is a word that means the opposite of another word or almost the opposite. Smile and frown have nothing in common.

To remember the difference between a synonym and an antonym, think of a synonym as the same (they both start with an s) and an antonym as the opposite or the antithesis.

On the ASVAB, you may be asked to find an equivalent, a word that most closely means the same thing as a given word. Or, you might be asked to find an antonym, which is the word that most closely means the opposite of a given word. Most of the questions on the Word Knowledge subtest will ask you to find words that mean the same thing but are different.

How can you study and find the synonym of a word (or the antonym, for that matter)? Take a look at these suggestions:

> » **Start in the dictionary.** Many dictionary entries include the abbreviation *syn.*, which means synonym. The words that follow this abbreviation are synonyms of the entry word. You may also see the abbreviation *ant.*, which stands for antonym; the word or words that follow it mean the opposite of the entry word.
> » **Make a list of synonyms and antonyms of the words you learn.** As you study vocabulary words for the Word Knowledge subtest, add them to your list.
> » **Use the root-word list from Table (earlier in the chapter).** Using a dictionary and/or thesaurus, develop a list of synonyms and antonyms for each word listed in the Sample Word column. (Of course, not every word has synonyms and antonyms, but many do.) Many ASVAB Word Knowledge questions require you to know a one-word definition for another word. There's no better study aid for this concept than a thesaurus, a book of Synonyms.

ASVAB Word Knowledge Strategy:
What to Do When You Don't Know the Answer:

Even if you don't know many words, you can still do well on the Word Knowledge subtest if you have a few tricks up your sleeve. Read this part and then use these strategies to answer the Word Knowledge questions at the end of this chapter and on the practice tests.

Making Your Situation:

Some Word Knowledge questions you'll see on the ASVAB don't have any context that can offer clues about their meaning (some are in sentences, which can make them easier to decode). The good news is that you may be able to give the word *your own* context, and when you do, you may find that you know the answer.

When you see a word you don't know, try to place it in context. Ask yourself, "Have I heard this word before?"

- Afoul most nearly means:

(A) correctly.
(B) wrongly.
(C) easily.
(D) disgusting.

Have you heard the word *afoul* in a sentence before? If you've heard of someone "running afoul of the law," you can surmise that the word doesn't mean correctly, easily, or disgusting—leaving you with Choice (B), which is the correct answer. It might be a fuzzy definition, but that's all you need sometimes.

Think of phrases you've heard that include these words (or variations), and then see whether you can guess their meanings.

- » **Abstain:** "We had abstinence-only education in school."
- » **Deduce:** "I'll try to deduce the answer based on what I know."
- » **Malignant:** "The tumor is malignant."
- » **Credible:** "Can you back that up with a credible source?"

(Abstain is a verb that means to restrain oneself from doing something. Deduce is a verb that means to conclude by reasoning. Malignant is an adjective that means life-threatening, spiteful, or mean. Credible is an adjective that means believable.)

What's it Like? Letting the Words Tell You What to do:

Words can be good or bad, and you can often tell which they are by what else is happening around them. So when a Word Knowledge question asks you to figure out the meaning of a word based on the sentence it's in, you can rule out wrong answers by how you feel.

- David hoped going to the amusement park would help him shake his melancholy.

(A) joyful
(B) sorrowful
(C) thoughtful
(D) excited

Most people have fun at an amusement park if it's not raining. It's almost as much fun as basic training, where "If it ain't raining, you ain't training" is often the day's motto. David also wants to "shake off his melancholy mod," meaning being sad can't be a good feeling. It must mean something bad (like that cold, prickly feeling). You can then eliminate Choices (A) and (D) and think about the opposite of fun. Most likely, it's Choice (B) because "sorrowful" means "down in the dumps."

Words and phrases like "dread," "looking forward to," and "shied away from" can help you figure out the tone of a sentence and choose the correct answer when you're not sure. Even getting rid of just one or two wrong answers can greatly affect how well you do.

Getting Clues: A Comparison and Difference:

Many sentence-based Word Knowledge questions have context clues that can help you decipher the underlined word's meaning. If you can pick up on the signal words that tell you about contrast and similarity, you can boost your score and vocabulary.

Signal words can be especially helpful in helping you predict a word's meaning.

- They recoiled as if they had just seen a ghost.

(A) cringed
(B) laughed
(C) shouted
(D) endured

In the question, the signal phrase is "as if," which is very close to "like." What's the most likely answer? If you saw a ghost, you probably wouldn't laugh, scream, or stand it.

(I'd run.) Choice (A), cringed, is the best answer because that's what you're most likely to do.

One of These is Not the Same as the Other: Getting Rid of the Wrong Answers:

One of the most successful ASVAB strategies involves ruling out answers that aren't likely correct before settling on the one you think is right. For example, when two answer choices mean almost the same thing, or when some choices don't match the prefix, suffix, or root word, you can cross them off as possibilities.

- Inhabit most nearly means:

(A) vacate.
(B) reside.
(C) depart.
(D) leave.

Choices (A), (C), and (D) are all very similar in meaning, leaving only Choice (B) to reside as the clear front-runner.

On the ASVAB, your choices aren't likely to be this obvious, but you may be able to rule out two choices and give yourself a 50-50 chance of finding the correct answer.

- Deform most nearly means:

(A) cure.
(B) heal.
(C) contort.
(D) tragedy.

Care to Fill in? Change the Word to One of the Choices:

When you get a Word Knowledge question that asks you to explain a word in a sentence, you might find that the right answer comes from swapping out the underlined word with each of your choices.

- The CDC hasn't been able to identify the pathogen that these people ingested before turning into zombies.

(A) person
(B) carrier

(C) event
(D) bacteria

Replace *pathogen* with *person, carrier, event,* and *bacteria.* (The word *ingested* is a big clue. The words *person, carrier,* and *event* don't make sense in the sentence.)

This strategy can work in conjunction with other strategies, too:

- Maria was worried that the cockroaches in the apartment would have a harmful effect on her daughter's health.

(A) helpful
(B) harmful
(C) erasing
(D) backward

Replace *deleterious* with each of the answer choices. Choice (B), *harmful,* is the only one that makes sense—especially because the sentence says, "Maria was worried," which shows you that the correct answer won't be something positive.

Parts of Speech:

When you see a Word Knowledge question that asks you to define a word in a sentence, you can weed out one or two incorrect choices by knowing the parts of speech. A suffix can tell you what part of speech a word is, so when you encounter a word you don't know, you can eliminate possible answer choices that you *do* know by looking at the suffixes.

Suffixes are usually only part of nouns, verbs, and adjectives. Therefore, if the ASVAB asks you to define *analogous,* but you don't know what it means, you can use your knowledge of the suffix to help you rule out answers that don't make sense. For example, because the suffix *-ous* typically modifies nouns and turns words into adjectives, you know that the correct answer probably won't be a noun.

- Analogous most nearly means:

(A) inclusive.
(B) danger.
(C) write.
(D) comparable.

Of those answer choices, *inclusive* and *comparable* are the only adjectives. *The danger* is a noun, and *write* a verb. You can rule out Choices (B) and (C) to make a more educated choice—and improve your chances of answering the question correctly—if you know that the suffix *-ous* usually refers to adjectives. Choice (D) is correct here; *analogous* means comparable in certain respects, especially when it clarifies the relationship between two things that are being compared ("The relationship between a drill sergeant and a recruit is analogous to the relationship between a hungry fish and a worm on a hook").

- Cameron knew the only viable option was investing 10 percent of his savings.

(A) succeed
(B) glowing
(C) reasonable
(D) life

The underlined word, viable, describes Cameron's option; option is a noun in this sentence, making viable an adjective. Look through the answer choices and figure out what part of speech each word is. Choice (A), succeed, is a verb because it describes an action (the action of succeeding). Choice (B), glowing, is an adjective because it modifies a noun (for example, "the glowing candle"), so that's a possible answer. Choice (C), reasonable, is also an adjective because it modifies nouns (as in, "That's a reasonable alternative"), so that's another option. Choice (D), life, is a noun (which wouldn't make any sense in this sentence), so it's off the table.

Choices (B) and (C) are the most likely of the four to be correct. If you haven't tried it yet, replace viable with each choice. You'll see that Choice (C), reasonable, makes the most sense in the sentence.

If Nothing Else Works, Break Your Word:

If you're stumped on a Word Knowledge question, you can break apart the word (read about that in "Deconstructing words" earlier in this chapter). Taking off a prefix or suffix can point you in the right direction.

- Predict the meanings of these words to test your skills:

» Desensitize
» Decode
» Deplume

These words all have something in common: the prefix *de-*. Knowing that *de-* usually means "away from," you can guess what each word will mean.

- Predict the meanings of these words:

» Approachable
» Serviceable
» Governable

The common theme with these words is the suffix *-able*, which typically signifies "capable of."

You can also use that to your advantage when taking the ASVAB.

Improve Your Vocabulary and Yourself:

If you want to do well on the Word Knowledge subtest, you should work on expanding your vocabulary. However, the strategies in this section can help you succeed even if you don't have an extensive vocabulary.

Vocabulary words can be learned quickly or over a longer period.

The optimum method is to combine both, but if you're short on time, memorizing key points and developing test-taking strategies are your best bets.

You Can Learn More Words by Reading:

Reading for pleasure is declining rapidly in today's environment of frequent social media updates and 17 billion streaming channels. Yet, reading is the single most effective way to expand one's vocabulary. According to studies, those who read for pleasure tend to have a greater vocabulary than those who don't.

The written word, as opposed to merely being told about it, is the words. Furthermore, there are far more words that people can read and understand than they can use in everyday discourse.

You don't have to jump into Advanced Astrophysics if you don't want to. If you don't often read very much, you can get started with something as simple as your daily newspaper or a news magazine. Focus on things that excite you. Reading more on the topic is fun if you're curious about it. To top it all off, you might pick up some helpful information!

The best way to determine the meaning of a term you don't know is to observe how it is used in context. If you come across the sentence "The scientist extrapolated from the data" but don't know what the word "extrapolated" means, you might try to make sense of it by replacing it with words you are familiar with. The scientist, for instance, probably didn't try to evade the data.

They probably used the information to form opinions or educated assumptions. Check the word's definition to be sure you're using it correctly. Making such forecasts can aid in the long-term retention of a definition.

Keep a running glossary of terms you encounter while reading and definitions (which you can find below). You won't always be able to figure out what a word means on the ASVAB's Word Knowledge subtest just by looking at it (frequently, there is no context in the test because the terms aren't used in sentences). You can forget about looking up the word in the dictionary, too. However, context clues and the dictionary are excellent resources when studying for a vocabulary test.

Making a List and Double-Checking it:

Not long ago, an 11-year-old girl went through the entire dictionary and listed all the words she didn't know. (The process took several months.) She then studied the list faithfully for a year and won first place in the National Spelling Bee finals. Of course, you don't have to go to this extent, but even putting in a tenth of her effort can dramatically improve your scores on the Word Knowledge subtest.

One way to improve your vocabulary is to keep a word list. Here's how that list works:

1. **When you hear or read a word you don't understand, jot it down or note it on your smartphone.**
2. **When you can, look up the word in the dictionary and then write the meaning on your list.**
3. **Use the word in a sentence that you make up.** Write the sentence down, too.
4. **Use your new word in everyday conversation.**

Finding a way to work the word *zenith* into a description of last night's basketball game requires creativity, but you won't forget what the word means. Arrange your list with related items so the words are easier to remember. For example, list the words related to your work on one page, words pertaining to mechanical knowledge on another page, and so on.

You can also find websites that offer lists of words if you spend a few minutes surfing. Try using search phrases such as "vocabulary words" and "SAT words."

Here are a few resources:

» **Vocabulary.com:** This site (www.vocabulary.com) offers thousands of vocabulary words and their definitions and interactive, adaptive games to help you learn.
» **Dictionary.com and Thesaurus.com:** Dictionary.com (www.dictionary.com) includes a great online dictionary and word of the day. The related site Thesaurus.com (www.thesaurus.com), which links back to the dictionary, gives you the same word of the day and lists of synonyms and antonyms.
» **Merriam-Webster online:** Merriam-Webster online (www.merriam-webster.com) is another useful site with a free online dictionary, thesaurus, and word of the day.

Crosswords are a Fun Way to Learn Words:

My grandmother always had a book of crossword puzzles and an ink pen in the middle of the kitchen table. (She must be pretty skilled if she writes her crossword puzzles!) So what, then, was her big secret? Her crossword-solving career began in the 1940s, long before anyone had access to a smartphone with a word game app.

In addition to being entertaining, the diversity of difficulty levels available in crossword puzzles is a significant plus. Begin with a puzzle whose level of complexity is appropriate for your current word knowledge level. You'll be a word-processing machine before you know it, and you'll have a blast doing it. You can avoid spending money at the supermarket checkout by downloading one of the dozens of free crossword apps available for cell phones.

Questions to Test Your Word Knowledge:

Each of the following Word Knowledge drills has an underlined word in the question's stem. Please select based on how well it describes the highlighted word in the question.

Take note of how each question is posed. In these cases, choose the answer that most closely matches the meaning of the highlighted word. Choose the word that is most opposed to the one given. The highlighted word is embedded in a sentence on other questions. If so, choose the answer that most closely resembles the meaning of the highlighted word when used in the sentence.

Simple Questions to Test Your Word Knowledge:

1. Estrange most nearly means
(A) sharp.
(B) small.
(C) alienate.
(D) shiny.

2. Momentous most nearly means
(A) significant.
(B) small.
(C) reality.
(D) postpone.

3. Pollute most nearly means
(A) eliminate.
(B) contaminate.
(C) clean.
(D) confuse.

4. The loving couple celebrated their fidelity over the years on their anniversary.
(A) complications
(B) faithfulness
(C) regrets
(D) treachery

5. The Army commander decided to avoid trouble by returning to headquarters.
(A) help
(B) complicate
(C) prevent
(D) attack

6. Authority most nearly means
(A) power
(B) disadvantage
(C) lack
(D) weakness

7. Terry assumed his friend would turn left, but instead, they turned right.
(A) misunderstood
(B) confused
(C) pretended
(D) supposed

8. The detective turned away from the gruesome sight.
(A) cartoonish
(B) pleasant
(C) man-made
(D) unpleasant

9. The conclusion most nearly means
(A) evidence
(B) end
(C) story
(D) influence

10. The driving instructor told the student to merge into the right lane.
(A) turn abruptly
(B) blend in
(C) ignore
(D) return

Medium Word Questions About Knowledge and Practice:

1. The college student met with the bursar to discuss tuition payment options.
(A) planner
(B) treasurer
(C) politician
(D) ghost

2. The mother chastised her child.
(A) comforted
(B) carried
(C) lectured
(D) supervised

3. We often wondered why Daniel lived in such an opulent apartment.
(A) run-down
(B) lavish
(C) far away
(D) hideous

4. Paul sent all of his friends a salutary message on the Internet.
(A) beneficial
(B) profane
(C) funny
(D) interesting

5. The word most opposite in meaning to reflect is
(A) ponder.
(B) consider.
(C) ignore.
(D) speculate.

6. The word most opposite in meaning to forthright is
(A) honest.
(B) polite.
(C) blunt.
(D) outspoken.

7. The witness was happy to corroborate the man's explanation.
(A) exaggerate

(B) destroy
(C) confirm
(D) question

8. Barry thought he could forge the paperwork without getting caught.
(A) trash
(B) destroy
(C) fabricate
(D) forget

9. Acclaim most nearly means:
(A) enthusiastic approval.
(B) religion.
(C) help.
(D) program.

10. Quell most nearly means:
(A) launch.
(B) support.
(C) enrich.
(D) suppress.

11. The kinetic energy was converted to electricity.
(A) regulated
(B) disbursed
(C) frantic
(D) moving

12. This year, unlike last year, the Paris fashion industry has decided to eschew short skirts and high heels.
(A) favor
(B) manufacture
(C) shun
(D) sell

13. The attendant charged Karen a nominal fee to park in the crowded lot.
(A) expensive
(B) insignificant
(C) large
(D) crazy

14. You can't circumvent the process; you must follow the procedure.
(A) connive
(B) regale
(C) overcome
(D) face

15. The general relinquished command and retired.
(A) gave up
(B) continued
(C) reclaimed
(D) asserted

16. Repugnant most nearly means:
(A) accessible
(B) political
(C) distasteful
(D) doglike

17. The new recruits had backpacks full of illicit materials.
(A) required
(B) common
(C) adult
(D) banned

18. If you work for the federal government, you may be entitled to a public transit subsidy.
(A) forfeiture
(B) grant
(C) write-off
(D) loss

19. Kai asked the salesperson to expedite delivery.
(A) explore
(B) entropy
(C) speed up
(D) reform

20. The nation faced heavy sanctions from the United States.
(A) artillery
(B) disapproval
(C) reward
(D) penalties

Questions to Test Your Knowledge of Hard Words:

1. Obtrude most nearly means:
(A) condition.
(B) absorb.
(C) prepare.
(D) impose.

2. Now that you've read through it once, it's time to recapitulate the Word Knowledge chapter.
(A) discuss
(B) summarize
(C) test
(D) reread

3. Clemency most nearly means:
(A) mercy.
(B) force.
(C) imprisonment.
(D) compliment.

4. Latent most nearly means:
(A) hidden.
(B) dull.
(C) pretentious.
(D) active.

5. The word most opposite in meaning to blame is:
(A) attribute.
(B) reprove.
(C) muster.
(D) exalt

6. Amenable most nearly means:
(A) amended.
(B) prepared.
(C) guided.
(D) cooperative.

7. Calamity most nearly means:
(A) desert.
(B) disaster.
(C) remorse.
(D) overeating.

8. The word most opposite in meaning to profound is:
(A) thorough.
(B) mild.
(C) absolute.
(D) intense.

9. The word most opposite in meaning to keen is:
(A) fervent.
(B) reluctant.
(C) zealous.
(D) appetent.

10. Truncate most nearly means:
(A) shorten.
(B) fixate.
(C) pretend.
(D) turn.

Answers and Breakdowns:

Use this answer key to score the Word Knowledge practice questions.

1. C.
Estrange means to alienate, Choice (C). Note that estrange is a verb, and the only answer choice that's also a verb is Choice (C); the others are adjectives.

2. A.
Momentous is an adjective that means very significant and is a choice (A).

3. B.
Pollute means to contaminate, Choice (B).

4. B.
Fidelity is a noun meaning faithfulness and loyalty in a relationship. (If you've heard the Marine Corps motto, "Semper Fidelis," which means always faithful, this one may have been easy for you to figure out.

5. C.
Avert is a verb meaning to turn something away or prevent it from happening.

6. A.
Authority is a noun that means the power or right to give orders and make decisions.

7. D.
Assume is a verb that means to suppose to be the case, even without proof.

8. D.
Gruesome is an adjective that means grisly; it causes horror or Repulsion.

9. B.
Conclusion is a noun that means the end of an event or process and a judgment or decision reached by reasoning.

10. B.
Merge is a verb that means to combine or to blend into something else.

11. B.
Bursar is similar to the word reimburse. The question gives context clues about tuition payment, and that should give you enough clues to select the correct answer, Choice (B).

12. C.
Chastised means disciplined or punished, so Choice (C) is the correct choice. Choices (A), (B), and (D) are unrelated.

13. B.
Opulent is an adjective that means wealthy, rich, or affluent. Choice (B) is the answer closest in meaning. The other choices are unrelated or opposite of the meaning.

14. A.
Salutary is an adjective meaning beneficial, so Choice (A) is correct. If you took Spanish in high school, you might remember that a related word, salud, relates to health and well-being, making Choice (A) a good guess.

15. C.
Reflect is a verb that means to think deeply about, demonstrate, or give back. Ignore would be the opposite of the meaning of the choices given.

16. B.
Forthright is an adjective meaning straightforward and honest, so of the choices given, polite is the closest to the opposite.

17. C.
Corroborate is a verb meaning to back up or confirm a story.

18. C.
Forge is a verb meaning to counterfeit or fabricate.

19. A.
Used as a noun, acclaim means a shout of approval, so the answer is Choice (A)

20. D.
Quell is a verb meaning to defeat or suppress.

21. D.
Kinetic is an adjective meaning relating to or characterized by motion.

22. C.
Eschew is a verb that means to avoid or keep away from. Therefore, choice (C) is the correct answer, and the other answers are unrelated.

23. B.
Nominal is an adjective meaning insignificant or lower than the actual or expected value.

24. C.
Circumvent is a verb that means to find a way around an obstacle or challenge.

25. A.
Relinquish is a verb that means to give up voluntarily.

26. C.
Repugnant is an adjective that means gross, extremely distasteful, or unacceptable.

27. D.
Illicit is an adjective that means forbidden by custom, rules, or law.

28. B.
Subsidy is a noun that means the grant of money from the government or a public body to help keep prices affordable or competitive.

29. C.
Expedite is a verb meaning to speed up or make something happen sooner.

30. D.
Sanction is a noun meaning a threatened penalty for breaking a rule or law.

31. D.
The correct answer is Choice (D). Obtrude means to intrude or to impose oneself on another. The other choices are unrelated.

32. B.
Recapitulate is a verb that means to summarize. It's also the more extended version of the word recap. The correct answer is Choice (B). Choice (A) is somewhat close, but Choice (B) is the closest in meaning.

33. A.
Clemency means forgiveness or leniency in punishing a person. Choice (A) is the correct answer. The other choices are unrelated. Knowing prefixes can be useful when determining the definitions of many words. For example, you may have heard the word inclement used to describe stormy, severe weather.
If you know that the prefix in- can mean not, you can conclude that clement is likely mild and gentle, traits related to mercy.

34. A.
Latent means present but not visible or noticeable, so Choice (A) is the correct answer.
Latent can also mean dormant, but none of the answer choices relates to that definition.

35. D.
Blame can be both a noun and a verb; as a verb, it means to condemn, place responsibility, or accuse. So, exalt is the opposite in meaning.

36. D.
Amenable is an adjective meaning willing and cooperative.

37. B.
Calamity is a noun meaning disaster.

38. B.
Profound is an adjective meaning thoughtful, intellectual, and intense, so mild is the opposite in meaning.

39. B.
Keen is an adjective meaning interested and enthusiastic, so reluctant is the opposite choice.

40. A.
Truncate is a verb meaning to shorten or abbreviate.

Chapter 8
Paragraph Comprehension (PC)

Conquering Paragraph Comprehension Questions: Global Questions

A common question type in the PC section asks you to identify a passage's main idea or theme, the author's purpose, or the passage's tone. Here's what a Global question stem might look like

- Which of the following is the passage's main idea?
- The author's tone in the passage above can be characterized.
- The purpose of the passage above is to
- The passage above is primarily concerned with

To find the main idea in a passage, you must be able to distinguish the values an author assigns to different statements. In each passage that asks you to determine a main idea, the author will use supporting details to establish a central idea.

Simple facts, other people's opinions, and background information can all operate as supporting details in a passage. Those details are then used to support an author's claim, often in the form of a strong opinion, a recommendation, a prediction, or a rebuttal to another person's position.

For example, imagine a passage with the following central idea: "Purchasing gold is a wise investment." This claim is not a fact but rather a recommendation. The author must support this claim by offering reasons why gold is a good investment.

Perhaps it's because "the amount of gold available worldwide is set to plateau," or maybe it's because "gold is the worldwide monetary standard." Those two statements, alone or together, help to explain why gold would make a wise investment. However, the recommendation that gold is a good investment does not explain why gold is the worldwide monetary standard.

So, one method for separating the main idea from the rest of the passage is to realize that the supporting details and the main idea answer two different types of questions, as you can see:

This:	Answers the question
Main idea:	What does the author believe?
Supporting details:	Why does the author believe what they believe?

Sometimes, the main idea will be evident when the author expresses a strong opinion. Words and phrases like *thus, therefore, I suggest,* and *I believe* all indicate that the author is providing a solid conclusion. Other times, a passage's main idea will be more challenging to identify. In those passages, be prepared to work harder to separate supporting statements from the main idea. Often, one of the tricks to zero in on the main idea of a passage is to pay attention to structural clues that indicate a contrast.

Words such as but, though, and yet provide subtle hints into an author's point of view: the statement after the contrast word frequently reflects the author's opinion, especially in passages where the author disagrees with someone else's opinion.

Finally, on Global questions, it is often easy to eliminate a few answer choices immediately. Wrong answer choices for Global questions usually do one of the following:

- They're too specific, dealing with just one small passage detail.
- They're too general, going beyond the scope of the passage.
- They're contradictory to the information presented in the passage.
- They're too extreme; that is, they distort the author's opinion by overstating it.

INFERENCE QUESTIONS:

Sometimes, the ASVAB will ask you questions about the passage to which the answers are not directly stated in the text but are implied by the passage.

Here are just a few of the ways an Inference question might be worded:

Which of the following is implied by the passage?

The author feels that It can be inferred from the passage that

As you can see, the wording in Inference questions rarely points you to a specific part of the passage, as Detail questions do. Inference questions won't necessarily ask you about the most important ideas to the author, as Global questions do.

Instead, Inference questions require you to consider multiple statements in a passage and determine what else the author must believe from them. Because many inferences can be made from a given text, predicting before considering the answer choices is unnecessary. Instead, paraphrase the passage and move through the answer choices, selecting the one supported by the text.

Because the correct answer to an Inference question is simply the one answer choice that is fully supported by what is stated in the passage, you can also eliminate wrong answer choices because they commit one of the following errors:

- They contradict the information in the passage.
- They bring in outside information that is not discussed in the passage.
- They distort the information presented in the passage.
- They make an extreme claim that is not entirely supported by the passage.

Vocabulary-in-Context Questions:

One final type of question in the PC section will ask you for the meaning of a word used in the paragraph. These questions, which we call Vocabulary-in-Context questions, are pretty straightforward: the correct answer will be a word that can replace the word in question without altering the sentence's meaning.

To handle this question type, focus on the word in the question stem while reading the passage. Then, predict an answer by defining the word as it is used in context. Finally, attack the answer choices by looking for a word that matches your prediction. Once you have selected an answer choice, reread the initial sentence with the answer choice in place of the vocabulary word to ensure the sentence's meaning is the same.

Paragraph Comprehension Practice Set 1:

1. Four years ago, the governor came into office seeking to change how politics were run in this state. Now, it appears she has been the victim of her ambitious political philosophy. Trying to do too much has given her a reputation as being pushy, and the backlash in the state has let her accomplish little. As a result, she may very well lose in her reelection bid.

The governor's approach to politics was:

(A) business as usual
(B) overly idealistic
(C) careless and sloppy
(D) influenced by his critics

2. Since its first official documentation by Sir George Everest in 1865, Mount Everest in Nepal has been the "Holy Grail" of mountaineers. Sir Edmund Hillary of New Zealand and Tenzing Norgay of Nepal were the first men to complete the ascent to the peak in May 1953. This feat won them international acclaim, not to mention knighthood for Hillary. But much less celebrated is a woman's first successful ascent of Everest, which did not occur until May 16, 1975.

Junko Tabei of Japan was the first woman to reach the summit of the world's most famous single peak. Stacy Allison of Portland was the first American woman to scale its heights successfully in 1988.

A member of the first successful Mount Everest expedition was:

(A) Sir George Everest
(B) Junko Tabei
(C) Sir Edmund Hillary
(D) Stacy Allison

3. It is without question a travesty that our children are no longer given healthy, nutritious food options for lunch in our public schools. Hamburgers, pizza, and chocolate give our kids bigger waistlines, but these junk foods are also helping teach them poor eating habits. We must change the mindset that any food is good and start offering students better meals at the same prices. Otherwise, a new generation of obese Americans is a given.

According to the passage, over the past few years, school lunches have gotten:

(A) more expensive
(B) more exotic
(C) healthier
(D) less nutritious

4. Families often choose to replace old furniture when just a minor amount of maintenance is needed. To fix a wobbly wooden chair, follow these easy steps. First, check the joints of the chair to see if the chair is structurally sound. There should be small dowel rods and a corner block to keep the chairs together. Next, use a ripping chisel to remove the corner block. Once the block is free from the chair, you should be able to glue the joints back together. Finally, once the glue looks dry, place the corner block back on the chair and gently mallet the block onto the dowels. You will have saved the chair and your hard-earned money in no time!

After removing the corner block, you should:

(A) mallet the dowels into place
(B) check the joints for damage
(C) glue the joints together
(D) replace the corner block

5. James felt the pulse of the crowd. There was a low murmur just under the house music. Backstage, his bandmates were tuning or drumming lightly on tabletops. In a few moments, the whole country would watch the band play. What a change from those dingy bars and clubs a few years ago. Maybe all the hard work had finally paid off. Looking down at his callused hands, he wondered if this might be the break they had been working so hard for.

The tone of this passage is one of:

(A) sadness
(B) anticipation
(C) anger
(D) ambivalence

6. The packaging of many popular foods is deceiving to consumers. Too often, the print is small and hard to read. And if you can read it, it's often confusing or intentionally vague. This is especially true on the nutrition label. The government should do something about the nutrition labels on food because the existing laws don't go far enough.

The author would probably support which of the following?

(A) magazine advertisements for cigarettes
(B) allergy information prominently listed on food labels
(C) the fine print on a contract
(D) food ads in the Sunday paper

7. In an age where we have pills for depression, dysfunction, and aggression, not to mention headaches, it is essential that we not forget that many drugs can have serious side effects. These may range from internal bleeding, vomiting, or soreness in the limbs to, in extreme cases, loss of consciousness or even coma. If you experience unwanted side effects, getting to a hospital immediately and seeking treatment is essential. Drinking alcohol or smoking cigarettes may also contribute to violent side effects.

According to the passage, one of the possible side effects of drugs is:

(A) drinking alcohol
(B) dysfunction
(C) internal bleeding
(D) aggression

8. First created at the height of postwar atomic paranoia, the Incredible Hulk stories offer a fascinating look at the dual nature of human beings. On the one hand, he is a mild-mannered, bespectacled scientist. On the other, he is a raging, rampaging beast. More than a statement about nuclear dangers, the Hulk reflects the two sides in each of us—the calm, logical human and the raging animal
According to the author, the comic book character of the Hulk is:

(A) a reflection of humanity
(B) an animal
(C) mild-mannered
(D) a protest about atomic power

9. Once considered the best high school player in the country, the onetime prodigy now spends his days working as a bricklayer for a local construction company. If he is bitter about how his life turned out, he replies, "Not at all." He says his only regret is that he didn't study hard enough and attend college. He still gets recognized occasionally, but an extra 80 pounds and bad knees keep him from reliving his former glory on the court.
The word *prodigy* in the passage most nearly means:

(A) depressed loner
(B) shy scientist
(C) gifted youngster
(D) bitter malcontent

10. Celebrated as one of the greatest film directors of the twentieth century, Alfred Hitchcock made his name by creating some of the most critically acclaimed suspense films ever. But while other directors of the genre seemed content telling stories of domestic intrigue, Hitchcock was not afraid to make films centering on unconventional subjects. One of his most celebrated films, *The Birds*, has no typical antagonist: the suspense comes not from a person out to do wrong but from nature itself.
The main idea of the passage above is that:

(A) Alfred Hitchcock is one of the greatest film directors ever
(B) most suspense films tell stories of domestic intrigue
(C) Alfred Hitchcock was not afraid to tell unconventional stories
(D) *The Birds* is scarier than traditional suspense films

Paragraph Comprehension Practice Set 2:

11. Bassoons are double-reeded wood instruments that produce a deep baritone sound. A bassoon's pitch can be altered by adjusting a curved tube called a bocal. The deeper the bocal is inserted into the bassoon, the higher the pitch. Conversely, pulling the bocal out of the instrument lowers the pitch.
It can be inferred from the passage that the bassoon:

(A) is a larger instrument than the oboe
(B) is the deepest-sounding wood instrument
(C) is the only instrument that utilizes a bocal
(D) is capable of a range of pitches

12. Investors who believe that a company's stock price is overvalued can perform an action called a "short sale." To short a stock, an investor sells shares she does not currently own; typically, shares for short selling are borrowed from a broker. If the stock drops in price, the investor can buy the shares at a lower price. Shorting stocks is risky, though, if the stock suddenly comes at a higher price, resulting in a significant loss.
According to the passage, an investor who shorts stocks first:

(A) buys shares of a company's stock, then sells those shares for a loss
(B) sells shares of a company's stock, then buys those shares at a later date
(C) buys shares of a company's stock, then sells those shares whenever he wishes
(D) sells shares of a company's stock, then uses that money to buy stock in another company

13. As rays of sunlight peeked through the tops of the trees, Mark felt at peace. It had been years since he had walked these trails. Everything was quiet and calm. It seemed to him that he finally saw the woods as they truly were: a refuge, a place to be renewed, a respite from the clamor and chaos of the city.
In this passage, *respite* most nearly means:

(A) relief
(B) opposite
(C) valley
(D) hope

14. The bus system in the city needs a drastic overhaul. Last week, the Main Street bus line was reduced from one bus every 15 minutes to one bus an hour. The week before that, the Uptown bus was canceled entirely. Each change occurred with no notice! The city expects people to continue to use public transportation; it needs to reduce the bus system issues.
Which of the following is mentioned in the passage?

(A) The city needs to notify customers of bus line interruptions.
(B) The city should allocate money from other programs to revamp the bus system.
(C) The Main Street bus line has experienced reduced service in the past.
(D) The Main Street bus line has been shut down in the past.

15. The European hedgehog is found in a wide range of habitats in Western Europe. It is popular because of its charming appearance and appetite for many pests plaguing European gardens. Unlike warmer-climate species of hedgehog, the European hedgehog is known to hibernate in the winter. It is a solitary animal, although occasionally a male and female will share a hibernation nest. While currently stable across much of continental Europe, the population of European hedgehogs is thought to be declining severely in Great Britain.
Which of the following best describes the topic of this passage?

(A) the decline of the European hedgehog in Great Britain
(B) why the European hedgehog is popular among European gardeners
(C) the characteristics and behavior of one species of hedgehog in Europe
(D) the hibernation habits of European hedgehogs

16. Questions 7 and 8 refer to the following paragraph. In 1918, as the world prepared to celebrate the end of World War I, a stealthier form of death appeared: the so-called "Spanish influenza." It is said that this strain of influenza killed more in a single year than the bubonic plague killed in a century. The outbreak gave modern scientists their first close look at a worldwide pandemic and paved the way for great medical advances. Furthermore, the unprecedented number of patients led to a boom in the medical field. One lasting result was increased pay for doctors, encouraging many to enter the profession. It could be said that the Spanish flu introduced the idea of "medicine for profit" to the world.
It can be inferred from the passage that:

(A) advances in medicine in the early twentieth century would not have happened without the outbreak of the Spanish flu
(B) scientists know little about the cause of the bubonic plague
(C) a doctor today might not have chosen to enter the medical profession before World War I
(D) the huge numbers of Spanish flu patients overwhelmed the available medical care and resulted in drafting more doctors to deal with the epidemic

17. In the context of the paragraph, the word *unprecedented* most nearly means:

(A) extraordinary
(B) tragic
(C) perplexing
(D) advantageous

18. Before the Great Chicago Fire of 1871 was extinguished, rumors circulated that it had been caused by the now infamous "Mrs. O'Leary's cow." Yet, after nine days of questioning fifty people, investigating authorities issued an inconclusive report about the fire's cause. "Whether it originated from a spark blown from a chimney on that windy night," the report read, "or was set on fire by human agency, we cannot determine." Besides Catherine O'Leary and her cow, suspects at the time included several colorful residents of Chicago's immigrant community, including "Peg Leg" Sullivan, who had first alerted the O'Leary family to the fire. In the end, the true cause or culprit will likely never be known.
Which of the following statements best expresses the main idea of the passage?

(A) Although the investigation was inconclusive, it is now considered likely that Catherine O'Leary's cow was the actual cause of the Chicago fire.
(B) Investigative authorities believed that the Chicago fire was set by an unknown suspect rather than being an accidental event.
(C) Investigators considered "Peg Leg" Sullivan a more likely suspect than Catherine O'Leary in causing the Chicago fire.
(D) The cause of the 1871 Chicago fire will probably never be Determined.

19. Despite his famous theories about how children are sexual beings who may develop into adults with unconscious psychological conflicts, Freud had only one patient during his lifetime who was a child. But this fact alone is not sufficient reason to dismiss his entire theory. Many of his ideas were based on information gained from reliable case histories of his adult patients. While some of his conclusions have been questioned in light of recent findings in neuropsychology, his most noteworthy contributions to psychology stand up against the criticisms of those who accuse him of fabricating his evidence to confirm his preconceived theories.
The author would most likely agree with which of the following?

(A) Neuropsychology has recently proven that Freud used unreliable evidence to support his theories.
(B) Although recent science has weakened some of Freud's theories, he should still be seen as making significant contributions to psychology.
(C) Freud's most famous theories are called into question because he based them all on one child patient.
(D) Freud's theories of child sexuality have been confirmed by numerous case histories.

20. At the beginning of the universe, temperatures were incredibly high. During this period of high energy, vast amounts of hydrogen, helium, and lithium were created. Although hydrogen and helium are still abundant, scientists believe the amount of lithium currently measured comprises only about a third of what we should expect. There are various explanations for why this might be, including some involving hypothetical elementary particles known as axions. Others believe that lithium is trapped in the core of stars, making it undetectable by our current instruments. Of the many theories proposed, there is no clear front-runner to explain the absence of lithium in the universe.
According to the passage, axions have been hypothesized to:

(A) show how lithium was created at the beginning of the universe
(B) explain why scientists detect less lithium in the universe than expected
(C) explain why hydrogen and helium make up most of the mass of the universe
(D) support the belief that lithium has become trapped in the core of stars

Answers and Explanations
PARAGRAPH COMPREHENSION PRACTICE SET 1:

1. **B**
From the information given, you should have seen that the governor was determined to get things done her way. So her approach could hardly be described as business as usual, choice (A). Nowhere in the passage is her approach described as a careless and sloppy choice (C). And since the mayor insisted on doing things her way, her approach was not influenced by her critics, choice (D). Only (B) is addressed in the paragraph. Seeking to change how politics are run indicates an idealistic approach to politics.

2. **C**
This Detail question asks you for a member of the first expedition to conquer Mount Everest. Researching the passage, you must be careful not to assume that the mountain is named for its conqueror. While Sir George Everest (A) first documented and recorded the height of Everest in 1865, it was (C) Sir Edmund Hillary who, along with Tenzing Norgay, completed the first ascent

to the peak in 1953. The question does not ask for the first woman or American woman to reach the summit, so (B) Junko Tabei and (D) Stacy Allison are out. The correct answer is (C).

3. **D**
The author feels strongly that meals in school cafeterias have become more and more similar to junk food. Choice (A) is not applicable, regardless of its validity, because it is not the central point of the passage. Choice (B) is nowhere indicated in the passage, and choice (C) is the opposite of the correct answer. Of the answer choices given, only choice (D), less nutritious, correctly answers the question.

4. **C**
This Detailed question asks you to identify a specific step in a process, so your first task is to locate the step that discusses removing the corner block. Removing the corner block is mentioned in the fifth sentence. After using a ripping chisel to remove the block from the chair, the worker can glue the joints back together to tighten them, choice (C).

5. **B**
To correctly gauge the tone of a passage, you should pay attention to the details, language, and description. In this passage, James is waiting to go onstage. As James does, words like pulse and murmur may invite the reader to feel excitement and nervousness. Of the answer choices given, (D), ambivalence is wrong, as James cares about what will happen. He seems wistful but never sad (A) or angry (C). The only answer that captures James's mood is choice (B), Anticipation.

6. **B**
Judging from the critical tone of the author and the subject matter, one can safely assume that the author is interested in public safety. Clearly, this author will not favor small print, so choice (C) is out. Choice (D) doesn't make sense, and choice (A) is not considered suitable for public safety. Choice (B), allergy information prominently listed on food labels, fits the author's passion for consumer labels.

7. **C**
According to the details of the passage, one possible side effect of prescription drug usage is internal bleeding, answer choice (C). Drinking alcohol (A) can exacerbate side effects, but it is not a side effect. Dysfunction (B) and aggression (D) are discussed as conditions treated by drugs, not as side effects of drug use.

8. **A**
From the details of the passage, it is clear that the author sees the character of the Hulk as a symbol. The passage states that the Hulk reflects the two sides of each of us, which matches choice. (A) It is a reflection of humanity. The passage states that the Hulk varies between an animal mentality (B) and a mild-mannered person (C), but it does not claim that he is one over the other. While the passage mentions "atomic postwar paranoia," it does not suggest that the Hulk is used as a protest of any type, as in choice (D).

9. **C**
According to the passage, the one-time star athlete is now a local bricklayer. There is nothing in the passage to indicate that the person in question was ever a depressed loner (A), shy scientist (B), or bitter malcontent (D), but it only makes sense that he was once a gifted youngster (C). The term "prodigy" refers to a person with natural ability, often at a young age.

10. **C**
Remember to look to contrast keywords to help determine an author's main point. Here, the passage makes a distinction between the work of typical directors and that of Hitchcock: Hitchcock, unlike the others, dared to make unconventional films, which is answer choice (C). While the passage supports answer choices (A) and (B), they are too narrow to encompass the main point. Answer choice (D) is out of scope and is not stated or implied in the passage.

PARAGRAPH COMPREHENSION PRACTICE SET 2:

11. **D**
While it may be true that the bassoon is larger than the oboe, that claim is not supported by information in the passage, so (A) is out. The passage states that the bassoon produces a deep sound, but claiming it is the deepest-sounding wood instrument is too extreme, so (B) is out. Nothing in the passage supports the claim that the bassoon is the only instrument that utilizes a bocal, which means (C) is incorrect. However, the passage does state that a bassoon's pitch can be altered by adjusting the bocal, so (D) is correct.

12. B

In this detailed question about a process, refer to the passage to find the correct answer. The information is in the second, third, and fourth sentences. The second sentence states that an investor sells shares she does not currently own, while the third and fourth sentences indicate that she purchases those shares later. That sequence is described in answer choice (B). Both (A) and (C) are incorrect because a short seller does not first buy shares of stock. (D) is out because although a short seller first sells stock shares, nothing in the passage indicates that that money is then used to purchase stock in another company.

13. A

Always predict Vocabulary-in-Context questions. The tone of this passage is one of peacefulness and calm. Therefore, you could infer that Mark's quiet, relaxing walk in the woods is a break, or a period of relief, from the clamor and chaos of the city. While (B) opposite might sound appealing since the passage contrasts the forest with the city, the opposite doesn't make sense if inserted into the sentence. Mark might be in a (C) valley, but that doesn't make sense if inserted into the sentence. Finally, (D) hope doesn't match the idea that Mark has found a quiet place away from the city.

14. C

This statement is a specific detail taken from the second sentence of the passage, "the Main Street bus line was reduced from one bus every 15 minutes to one bus an hour." There is no evidence that the Main Street bus line has been completely shut down (D). The author does mention not being given any notification, but he does not state choice(A), that the city needs to notify customers. Choice (B) might be an idea the author would agree with, but it is not mentioned in the passage.

15. C

The author introduces the European hedgehog as "a species" of hedgehog in the first sentence, comparing it to other species. The passage then provides information about the European hedgehog's eating and hibernation habits and the males' aggressive behavior. Choice (C) captures the paragraph's focus and correctly answers this Global question. Choices (A), (B), and (D) all mention facts given in the passage but miss the big picture by narrowly focusing on only one Idea.

16. C

The question asks for an inference that can be drawn from the passage but provides no clues about the nature of the inference. Each answer choice must be compared with the information given to identify the one choice that must be true based on the passage. The fifth sentence supports the correct answer: "One lasting result was an increase in pay for doctors, encouraging many to enter the profession." If many people were encouraged to become doctors because of higher pay, we could infer that lower pay before World War I was a deterrent to entering the profession. The word "lasting" implies that higher pay still motivates more people to become doctors today. Choice (A) is extreme; even though the flu brought about advances, you cannot infer that advances would not have happened. Choice (B) is unsupported by any facts presented. Choice (D) distorts the facts given. Doctors were encouraged by higher pay, not drafted. Also, there is no evidence that medical care was overwhelmed by the unprecedented numbers of patients. The patients caused a "boom" in the field.

17. A

This Vocabulary-in-Context question asks about the author's use of the word unprecedented. The author uses the word to describe the increase in the number of patients, which had a lasting impact on the medical profession. The number of patients must have been unusual or extraordinary to have such a significant effect. Choice (B), tragic, might be tempting because of the number of people who died. But this meaning does not fit the point of the sentence. Choice (C) can be eliminated because the text does not suggest that the number of patients was perplexing or puzzling. Choice (D), advantageous, would indicate that the author thinks the number of patients was beneficial, which is not the author's point. The change in the medical profession may or may not have been advantageous; the author gives no opinion.

18. D

This Global question asks for the author's main idea. The word "Yet" in the second sentence signals the author's key point: authorities could not establish the cause of the fire. The final sentence reinforces this point. The rest of the text discusses the fire investigation and speculation about its causes. Thus, choice (D) is correct. Choice (A) is never stated or implied in the passage. Choice (B) contradicts the report's quote, which says whether the cause was accidental or intentional is unknown. Choice (C) is a distortion; "Peg Leg" Sullivan was considered a suspect, but no likelihood of his guilt is proposed.

19. B

The correct answer to this Inference question will reflect an opinion given by the author in the text. The first sentence points out a perceived problem with the data supporting one of Freud's theories. The author then states in the second sentence that this weakness in the data is not enough to dismiss Freud's ideas. The rest of the paragraph supports this claim, concluding that Freud's contributions to psychology still stand up. Choice (B) sums up the author's opinion. Choice (A) brings up an idea presented, but the contrast word "While" in the passage indicates that this is not the author's opinion. Choice (C) brings up the problem with Freud's evidence mentioned in the first sentence. This, however, is the point that the author dismisses in the second sentence. Choice (D) distorts the author's defense of Freud; the word "confirmed" is extreme and does not match the author's concession that some of Freud's conclusions have been Questioned.

20. B

This Detail question asks about something specifically mentioned in the passage. Axions are mentioned in the fourth sentence, which begins, "There are a wide variety of explanations for why this might be . . ." to introduce one proposed explanation for the unexpected scarcity of lithium in the universe. Choices (A) and (C) both refer to the creation of various elements, and the author does not mention axions in relation to how the elements came into being. Choice (D) distorts the idea of axions by joining it with another explanation, set apart in the following sentence by the words "Others believe."

Chapter 9

Mathematics Knowledge (MK)

A purported letter from Albert Einstein to a fan read: "Do not worry about your problems with mathematics." Mine are much bigger. So, it's safe to assume that the respected professor never had to sweat the ASVAB. However, a strong performance in the Mathematics Knowledge section does not require theoretical expertise in mathematics. Therefore, no advanced education in mathematics is assumed for this section of the test.

The Mathematics Knowledge section of the ASVAB computerized test has 15 questions and takes 23 minutes to complete. The 25 questions on the paper and pencil ASVAB must be answered in 24 minutes. If you get sample questions for this exam section, you'll have 47 minutes to complete 30 questions. You don't have to rush through every calculation, but you also can't afford to take your time.

If you're taking the P&P version of the exam, you can move on to the next question if you're unsure of the answer (ensure you fill in the correct fields on your response sheet). Concentration and focus are required for speedy and precise problem-solving. In addition, calculators are not permitted. However, test proctors provide you with as much scratch paper and fresh pencils.

Because each question on the CAT-ASVAB must be answered before going on to the next, skipping questions is not an option. Remember that your experience with the CAT-ASVAB may be more challenging (or easier) than anticipated.

Because the software adjusts to your skill level, it starts with a question of medium difficulty and then adjusts the difficulty of subsequent questions based on how well you did on the initial question. So if you're in the thick of the test and thinking, "This is harder than what the book covered," you're probably doing quite well.

There are often only a few questions in the Mathematics Knowledge section of the exam—each specific mathematical concept. For example, one question may ask you to multiply fractions, the next may ask you to solve a mathematical inequality, and the question after that may ask you to find the value of an exponent. (If the last sentence freaks you out, calm down. I cover each of these concepts in this book.)

All this variety forces you to shift your mental gears quickly to deal with different concepts. You can look at this situation from two perspectives. These mental gymnastics can be difficult and frustrating, especially if you know everything about solving for x but nothing about finding a square root. But variety can also help save you on the test if you don't know how to solve a specific problem. That oversight may cause you to get only one or two questions wrong.

Math has its vocabulary. To understand what each problem on the Mathematical Knowledge subtest asks, you need to understand specific mathematical terms:

» **Integer:** An *integer* is any positive or negative whole number or zero. The ASVAB often requires you to work with integers such as –6, 0, or 27.

» **Numerical factors:** Factors are integers (whole numbers) that can be divided evenly into other integers. To *factor* a number, you determine the numbers you can divide into it. For example, 8 can be divided by the numbers 2 and 4 (in addition to 1 and 8), so 2 and 4 are factors of 8. The *prime factorization* of the number 30 is written 2 3 5. Numbers may be either composite or prime, depending on how many factors they have:

- **Composite number:** A *composite* number is a whole number that can be divided evenly by itself and by 1, as well as by one or more other whole numbers; in other words, it has more than two factors. Examples of composite numbers are 6 (whose factors are 1, 2, 3, and 6), 9 (whose factors are 1, 3, and 9), and 12 (whose factors are 1, 2, 3, 4, 6, and 12).
- **Prime number:** A *prime* number is a whole number that can be divided evenly by itself and by 1 but not by any other number, meaning it has exactly two factors. The number 1 is not a prime number. Examples of prime numbers are 2 (whose factors are 1 and 2), 5 (whose factors are 1 and 5), and 11 (whose factors are 1 and 11).

» **Base:** A *base* is a number used as a factor a specific number of times—a number raised to an exponent. For instance, the term 43 (which can be written 4 4 4, and in which 4 is a factor three times) has a base of 4.

» **Exponent:** An exponent is a shorthand method of indicating repeated multiplication. For example, 15 15 can also be expressed as 15^2, also known as "15 squared" or "15 to the second power." The small number written slightly above and to the right of a number is the exponent, and it indicates the number of times you multiply the base by itself. Note that 15^2 (15 15), which equals 225, isn't the same as 15 2 (which equals 30). To express 15 15 15 using this shorthand method, write it as 15^3, which is also called "15 cubed" or "15 to the third power." Again, 15^3 (which equals 3,375) isn't the same as 15 3 (which equals 45).

» **Square root:** The square root of a number is the number that, when multiplied by itself (in other words, squared), equals the original number. For example, the square root of 36 is 6. If you square 6 or multiply it by itself, you produce 36. (Check out "Getting to the Root of the Problem" later in this chapter.)

» **Factorial:** A factorial is an operation represented by an exclamation point (!). You calculate a factorial by finding the product of (multiplying) a whole number and all the whole numbers less than it down to 1. That means 6 factorial (6!) is 6 5 4 3 2 1 720. A factorial helps you determine permutations—all possible ways an event may turn out. For example, if you want to know how many different ways six runners could finish a race (permutation), you would solve for 6!: 6 5 4 3 2 1.

» **Reciprocal:** A reciprocal is a number by which another number can be multiplied to produce 1; if you have a fraction, its reciprocal is that fraction turned upside down. For example, the reciprocal of 3 is ⅓. If you multiply 3 times ⅓, you get 1.

» **Rounding:** Rounding is limiting a number to a certain number of the innermost parentheses in cases where parentheses are contained within g operations all the time—often without even thinking about it. If you have $1.97 in change in your pocket, you may say, "I have about two dollars." The rounding process simplifies mathematical operations. Often, numbers are rounded to the nearest tenth. The ASVAB may ask you to do this. If the number you're eliminating is 5 or over, round up; for any number under 5, round down. For example, 1.55 rounded to the nearest tenth can be rounded up to 1.6, and 1.34 can be rounded down to 1.3. Many math problems require rounding—especially when doing all this without a calculator. For example, (π) pi represents a number approximately equal to 3.1415926535897932384626433 83 (and on and on). However, it's common to round to 3.14 in mathematical operations and on the ASVAB.

Operations: What You Do to Numbers:

When you toss numbers together (mathematically speaking), you operate. When you add or multiply, you perform an essential *operation*. But because math functions according to yin-yang-like principles, each of these basic operations also has an opposite operation called an *inverse operation*. Thus, the additive inverse (of addition) is subtraction, and the multiplicative inverse (of multiplication) is division. And, of course, the inverse of subtraction is—you got it—addition. The inverse of division is multiplication.

First Things First: Following the Order of Operations:

Operations must be performed in a certain order. For example, when you have parentheses in a math problem, you must do the calculation in the parentheses before you do any calculations outside of the parentheses. In the equation $2 \times (16 + 5) = ?$, you first add 16 to 5 to arrive at 21, and then you multiply by 2 to come up with a total of 42. You get a different (and wrong) answer if you simply calculate from left to right: $2 \times 16 = 32$, and $32 + 5 = 37$. And you better believe that both results will be choices on the test!

1. Parentheses Take Precedence.

You should do everything contained in parentheses first. Do the innermost parentheses in cases where parentheses are contained within parentheses first. This rule also goes for brackets and braces.

Note: If you're dealing with a fraction, treat the top as though it were in parentheses and the bottom as though it were in parentheses, even if the parentheses aren't written in the original state. Suppose you have the problem 3/1+2. Add the numbers below the fraction bar before dividing. The answer is 3/3=1 (For more on fractions, see the later section "Working on Both Sides of the Line: Fractions.") The square root sign(√) is also a grouping symbol, so you solve for whatever's under the top bar of the square root sign before doing any other operation in the problem.

2. Exponents Come Next.
Remember that the exponent goes with the number or variable that it's closest to. If it's closest to a parenthesis, then you already should've performed the calculation inside the parentheses in Step 1. For example, $(5x2)^2 = 10^2 = 100$. The square root sign $\sqrt{}$ is also treated as an exponent, so you take the square root during this step.

3. Multiplication and Division are the following.
Do these operations in left-to-right order (just like you read).

4. Addition and Subtraction are Last.
Perform these operations from left to right as well.

Check out the following example for a bit of practice with an order of operations:

(15/5)x3+(18-7)=?
Do the work in parentheses and then remove the parentheses:
3x3+11=?
No exponents are present, so division and multiplication come next (in this problem, only multiplication is needed):
9+11=?
Finally, do the addition and subtraction (in this problem, only addition is needed). Your final answer is 20.

Use the acronym PEMDAS, or "Please Excuse My Dear Aunt Sally," to remember the order of operations. It stands for "parentheses, exponents, multiplication, division, addition, subtraction" to show you which calculations to perform in order.

Check out these order of operations practice problems and see what you come up with.

Problem	Solution
7/(5x2-(6/3))=?	
7(4+13)=?	
16-5+(7-3)x3=?	
23x(8x4)/5=?	

The answers, in order from top to bottom, are 0.875, 119, 23, and 147.2.

Completing a Number Sequence:

The Arithmetic Reasoning (AR) subtest often includes questions that test your ability to name what comes next in a sequence of numbers. Generally, these problems are the only AR questions that aren't word problems (which I cover in Chapter 9).

However, sequence questions test your ability to do arithmetic and reason because you must determine how the numbers relate. To do this, you must be able to perform mathematical operations quickly. For example, suppose you have a sequence of numbers that looks like this: 1, 4, 7, 10, etc. Each new number is reached by adding 3 to the previous number: 1+3 =4, 4+3=7, and so on. So the following number in the sequence is 10+3=13 or 13.

But, of course, the questions on the ASVAB aren't quite this simple. You'll likely see something like this: 2, 4, 16, 256, ?. In this case, each number is being multiplied by itself, so 2x2=4, 4x4=16, and so on. The next number in the sequence is 256 256, which equals 65,536—the correct answer.

You may also see sequences like this: 1, 2, 3, 6, 12, ?. In this sequence, the numbers are being added together: 1+2=3, and 1+2+3=6. So the next number is 1+2+3+6=12. So the following number would be 24.

Finding the Pattern:

To answer sequence questions correctly, you need to figure out the pattern as quickly as possible. Some people, blessed with superior sequencing genes, can figure out patterns instinctively. But, unfortunately, the rest of the population has to rely on a more arduous, manual effort.

Finding a pattern in a sequence of numbers requires thinking about how numbers work.

For instance, seeing the number 256 after 2, 4, 16 should alert you that multiplication is the operation because 256 is much larger than the other numbers. On the other hand, because the values in 1, 2, 3, 6, and 12 don't increase by much, you can guess that the pattern requires addition.

Dealing with More than One Operation in a Sequence:

Don't forget that more than one operation can occur in a sequence. For example, a sequence may be "add 1, subtract 1, add 2, subtract 2." That would look like this: 2, 3, 2, 4, ?.

Because the numbers in the sequence increase and decrease as the sequence continues, you should suspect something tricky is going on. Make sure to use your scratch paper! Jot down notes while trying to find the pattern in a sequence. Writing down your work helps you track which operations you've been attempting. Try your hand at predicting the following numbers in the following sequence problems.

Problem	Solution
6,13,27,55,.....	
5,16,27,38,....	
8,14,22,36,58,....	
14/4, 4/2, 2/2, ½,....	

The answers are, in order from top to bottom, 111, 49, 94, and ¼

Averaging Mean, Median, and Mode in a Range:

An average number expresses a typical value in a set of data. In plain English, that means what you can expect as an outcome.

For example, if you take the ASVAB three times and get an AFQT score of 80, 85, and 90, respectively, your average score is 85. (I advise against retaking the ASVAB to practice your averaging skills; I want you to take it once and score well!) Things get a little sticky when the numbers aren't so nicely rounded. If someone's scores are 35, 35, and 50, their average is 40.

This average type is technically called the *arithmetic mean* or *mean* for short. To figure out a mean, add all the numbers in a data set (in this case, that's 35, 35, and 50). Then, divide the sum by how many numbers there are.

Because there are three numbers in this data set, your math looks like this:

35+35+50=120

120/3=40

You might also counter other averages on the ASVAB, including *median* and *mode*.

» **Median:** The median is the middle value in a set of ordered numbers. You can find it by putting your numbers numerically, from smallest to largest. In the data sets 47, 56, 58, 63, and 100, the median is 58.
» **Mode:** Mode is the value or values that occur most often in a list of numbers. If no numbers are repeated, there's no mode. But in the earlier data set with the test scores—35, 35, and 50—the mode is 35 because it appears most often.

Working on Both Sides of the Line: Fractions

I don't know why, but almost all math textbooks seem to explain fractions in terms of pies. (I think most mathematicians must have a sweet tooth.) But I like pizza, so I'm going to use pizza instead. If a *whole number* is a pizza, a *fraction* is a slice of pizza. A fraction also illustrates the slice's relationship to the whole pizza. For example, consider the fraction ⅗.

If you accuse your cousin of eating 3/5 of the pizza when they come over for movie night, you're saying that the pizza is divided into five equal-sized slices—fifths—and your cousin ate three of those five. The number above the fraction bar—the three slices your cousin ate—is called the numerator. The number written below the fraction bar—the total number of slices the pizza is divided into—is called the denominator.

Common Denominators: Preparing to Add and Subtract Fractions:

The fractions must have the same denominator (bottom number), known as a common denominator, to be added and subtracted. You must identify a common denominator if the fractions don't already have one. Two fundamental approaches can be used. A good time? Continue reading.

Method One

Finding a common denominator can be easy, or it can be as tricky as picking off anchovies.

Suppose you want to add 3/5 and . Getting a common denominator is easy here, and you use this process whenever you can evenly divide one denominator by another.

Follow these steps:

1. Divide the larger denominator by the smaller denominator.
If there's a remainder, you can't use this method, and you must use method two (see the next section). In this case, 10 can be divided evenly by 5. The quotient (answer) that results is 2.

2. Take the fraction with the smaller denominator 3/5; multiply the numerator (top number) and the denominator (bottom number) by the answer resulting in Step 1.
Multiply 3 by 2; the result is 6—that's your new numerator. Multiply 5 by 2; the result is 10—that's your new denominator.

3. Replace the numerator and denominator with the numbers from Step 2.
You can also express 3/5 as 6/10. (If you cut the pizza into 10 slices instead of 5 and your cousin eats 6 slices instead of 3, they've eaten exactly the same amount of pizza.)

After finding a common denominator, you add the two fractions by adding the numerators together: 6/10 + 3/10 =9/10. Think of it this way: If your cousin eats 6/10 of the pizza (which is just another way of saying 3/5) and you eat 3/10 of the pizza, together you've eaten 9/10.

Method Two

Suppose your cousin eats 3/5 of one pizza, and your sister eats 6 1 of another pizza (one cut into 6 slices instead of 5), and you want to know how much pizza has been eaten. In this case, you need to add 3/5 and 6 1.

Adding these fractions is a bit more difficult because you can't divide either denominator by the other. Instead, you must find a common denominator that both 5 and 6 divide evenly. Here's how:

1. Multiply the first fraction's denominator by the second fraction's denominator.
In the preceding example, 5x6=30. The common denominator for both fractions is 30.

2. Express the first fraction in terms of the new common denominator.
3/5 = ?/30

3. Multiply the numerator by the number you multiplied to produce the new denominator.
To convert the denominator (5) to 30, multiply by 6, so multiply the numerator (3) by 6. The result is 18. Therefore, the fraction can be expressed as 18/30.

When trying to find an equivalent fraction for a fraction, always multiply the numerator and the denominator by the same number. Otherwise, you change the value of the fraction.

With this example, you multiply the numerator and the denominator by 6, discovering that ⅗ is the same thing as 18/30. But if you multiply only the denominator by 6, you'd have a new number; ⅗ and 3/30 don't have the same value.

4. Next, express the second fraction

Using the new common denominator.
⅙ = ?/30

5. Multiply the numerator of the second fraction by the number you used to result in the denominator.

To get 30, you have to multiply 6 by 5. Multiply the numerator by the same number: You find that 1x5=5, so the fraction 61 can be expressed as 5/30. After all that work, you can finally add the fractions: 18/30 + 5/30 = 23/30. Now pause and take a bite of pizza.

Finding common denominators for three or more fractions

If you have more than two fractions with different denominators, you must find a common denominator into which all the denominators divide. Suppose you need to add ½ + ⅔ + ⅗

A simple way to find a common denominator is to take the largest denominator (in this case, 5) and multiply it by whole numbers, starting with 1, 2, 3, 4, and so on, until you find a denominator that the other denominators also divide into evenly.

If you multiply 5 by 2, you get 10, but 3 isn't divided evenly into 10. So keep going: 5x3=15, 5x4=20, and so on until you find a number that 2, 3, and 5 can divide into evenly. In this case, 30 is the first number you can find that 2, 3, and 5 can divide evenly, so 30 is your common Denominator.

Multiplying and Reducing Fractions:

Multiplying fractions is easy. You multiply the numerators and then multiply the denominators. So look at the following equation: ½ x ¾ x ⅗ =? So you multiply 1x3x3=9 (the numerators) and 2x4x5=40 (the denominators) to result in 9/40.

Occasionally, when you multiply fractions, you end up with an extremely large fraction that can be simplified or reduced. To express a fraction in its *lowest terms* means to put it so you can't evenly divide the numerator and the denominator by the same number (other than 1).

A common factor is a number you can divide into the numerator and the denomina*tor*. If you have the fraction 6/10, both the numerator (6) and the denominator (10) can be divided by the same number, 2. If you do the division, 6 2 3 and 10 2 5, you find that 6/10 can be expressed in the simpler terms of ⅗. You can't reduce (simplify) ⅗ any further; the only other number that the numerator and denominator can be divided by is 1, so the result would be the same, ⅗.

Remember, you can't use a calculator on the ASVAB, so multiplying large numbers can take extra steps and valuable time. You can make your work easier by *canceling out* common factors before multiplying.

For example, suppose you have the problem $\frac{20}{21} \times \frac{14}{25}$. Multiplying the numerators (20x14=280), then multiplying the denominators 21x25=525, and finally reducing the fraction (280/525 = 8/15) may require you to write out three or more separate multiplication/division problems. But you can save time if a numerator and denominator have common factors. Here, the numerator of the first fraction (20) and the denominator of the second (25) have a common factor of 5, so you can divide both of those numbers by 5: Your problem becomes 4/21 x 14/5 The numerator of the second fraction (14), and the denominator of the first fraction (21) are both divisible by 7, so you can cancel out a 7: Divide 14 and 21 by 7. This changes the equation to **4/3 x ⅖ =8/15**, a simpler math problem.

Dividing Fractions:

Dividing fractions is simple if you remember this rule: Dividing a fraction by a number is the same as multiplying it by the multiplicative inverse (reciprocal) of that number. Of course, there are always exceptions. You can't use this operation on zero. Zero has no multiplicative inverse. Instead, you obtain the reciprocal of a number by flipping the number.

If you want to divide a fraction by 5, you multiply the fraction by the inverse of 5, which is ⅕. You can understand this process more easily if you remember that 5 is the same thing as ⅕. In other words, 5 divided by 1 equals 5 (5/1=5). And the reciprocal of 5/1 is ⅕. To come up with the reciprocal of a number, stand the number on its head. To divide a fraction, use the inverse of the number that follows the division symbol ( ) and substitute a multiplication symbol (x) for the division symbol. Therefore, ⅓ /2 is expressed as $\frac{1}{3} \times \frac{1}{2}$, and you already know how to multiply fractions. 1x1=1 and 3x2=6, so the product of ⅓ x ½ = ⅙ Therefore, ⅓ /2 = ⅙

Converting Improper Fractions to Mixed Numbers and Back Again:

You have an *improper fraction* if you have a fraction with a numerator larger than or equal to its denominator. For example, 7/3 is an improper fraction. To put an improper fraction into simpler (proper) terms, you can change 7/3 into a *mixed number* (including a whole number and a fraction).

Divide the numerator by the denominator: 7 divided by 3 gives you a quotient of 2 with a remainder of 1. However, there's something left over because 3 doesn't divide evenly into 7. The remainder becomes a numerator over the original denominator, so 13 is left over. Therefore, 7/3 is the same as 2 ⅓.

If you want to multiply or divide a mixed number, you need to convert it into a fraction—an improper fraction. To make the change, convert the whole number into a fraction and add it to the fraction you already have.

Here's how:

1. Multiply the whole number by the denominator (bottom number) of the existing fraction to arrive at a new numerator.
Suppose you have
7 ⅔ . Multiply 7 by 3: 7x3=21.

2. Place this new numerator over the existing denominator.
21/3

3. Add that fraction to the original fraction for the final answer.

$$\frac{21}{3} + \frac{2}{3} = \frac{23}{3}$$

Check out the "Common denominators: Preparing to add and subtract fractions" section for the complete scoop on adding fractions earlier in this chapter.

Or, if you want to get technical, you can look at the whole process this way, too:

$$7\frac{2}{3} = \frac{7 \times 3 + 2}{3} = \frac{23}{3}$$

Expressing a Fraction in Other Forms: Decimals and Percents

A fraction can also be expressed as a decimal and as a percent. Here's how to convert between forms:

» **To change a fraction into a decimal:** Divide the numerator (top number) by the denominator (bottom number). Given that handy explanation, ⅗ converted into decimal form is 0.6.
Some fractions convert to *repeating decimals*—a decimal in which one digit is repeated infinitely. For instance,⅔ is the same as 0.66666 (with the sixes never stopping).
Repeating decimals are often rounded to the nearest hundredth; therefore, ⅔ rounds to 0.67.

(**Remember:** The first space to the right of the decimal is the tenth place, the second space is the hundredths place, the third is the thousandths, and so on.)

» **To make a decimal into a percent:** Move the decimal point two spaces to the right and add a percent sign. For example, 0.6 becomes 60%.

Using Ratios to Show Comparisons:

A *ratio* shows a relationship between two things. For example, if Margaret invested in her tattoo parlor at a 2:1 (or 2 to 1) ratio to her business partner, Julie, then Margaret put in $2 for every $1 that Julie put in. You can express a ratio as a fraction, so 2:1 is the same as 2/1.

Or suppose you fill up your shiny, brand-new SUV and want to compute your gas mileage—miles per gallon. You drive for 240 miles and then refill the tank with 15 gallons of gas, so the ratio of miles to gallons is 240:15. You can compute your gas mileage by dividing the miles by the number of gallons: 240 miles 15. So you're getting 16 miles per gallon. It's time for a tune-up!

Navigating Scale Drawings

Scale, mainly when used on the ASVAB, relates to scale drawings. For example, a map drawn to scale may have a 1-inch drawing of a road representing 1 mile of physical road in the real world. The Arithmetic Reasoning portion of the ASVAB often asks you to calculate a problem based on scale, representing a standard ratio (1 inch:1 mile) or a fraction (1 inch/1 mile).

On a map with a scale of 1 inch to 1 mile, the ratio of the scale is represented as 1:1. But questions are never this easy on the ASVAB. You're more likely to see something like, "If a map has a scale of 1 inch to every 4 miles . . ." That scale is expressed as the ratio 1:4, or 4/1

Try your hand with the following common scale problem: If the scale on a road map is 1 inch 250 miles, how many inches would represent 1,250 miles?

The problem wants you to determine how many inches on the map represent 1,250 miles if 1 inch equals 250 miles. You know that 1 inch is 250 miles, and you also know that some unknown number of inches, which you can call *x*, equals 1,250 miles. The problem can be expressed as two ratios set equal to each other, known as a proportion:

1/250 = x/1250 Now all you have to do is solve for *x* by multiplying each side of the equation by 1,250:

$$1/250 = x/1250$$
$$1/250 \times 1250 = x/1250 \times 1250$$
$$1250/250 = x$$
$$x = 5$$

So, if 1 inch is equal to 250 miles, then 5 inches would be equal to 1,250 miles. If this problem causes you to scratch your head, check out Chapter 7 for info on solving for *x*.

Almost every military job uses scales, which is why scale-related questions are so common on the ASVAB. Whether reading maps at Mountain Warfare School or organizing trash pickup (what the military calls "area beautification") around the base, you must use and interpret scales frequently.

Playing with Proportions

Proportion is a mathematical comparison between two ratios. So, for example, when you see math (or real-world) problems that say things like "3 parts vinegar, 1 part water," you're dealing with a ratio (and in this case, it's 3:1). But when you compare two ratios (3 parts vinegar, 1 part water and 9 parts vinegar, 3 parts water), you're dealing with a proportion.

If two sets of numbers increase or decrease in the same ratio, they're directly proportional. You can identify proportions by the symbol: or $\propto$ (just like the vinegar and water recipe I describe in the previous paragraph). You use direct proportions when an ASVAB question asks you to determine how much to increase or decrease a recipe, how long one vehicle takes to travel a certain distance at a certain speed, and a few other scenarios.

Inverse proportions occur when one value increases and the other decreases. For example, more workers on a job reduce the time necessary to complete a task. Therefore, you may see inverse proportions on ASVAB questions about multiple people filling up a swimming pool, time to complete a job, or something similar.

Writing in Scientific Notation:

Scientific notation is a compact format for writing very large or small numbers. Although it's most often used in scientific fields, you may find a question or two on the Mathematics Knowledge subtest of the ASVAB asking you to convert a number to or from scientific notation.

Scientific notation separates a number into two parts: a number between 1 and 10 and a power of ten (such as 10^7, 10^{21}, or 10^{-18}; see the earlier section "Just When You Thought You Were Done with Vocab: Math Terminology" for info on powers and exponents). Therefore, 1.25×10^4 means 1.25x10 to the fourth power, or 12,500; 5.79×10^8 means 5.79/10 to the eighth power, or 0.0000000579. The exponent tells you how many places to move the decimal point and whether to move it left (if it's negative) or right (if it's positive).

Finding the Problem's Source:
A *square root* is the factor of a number that produces the number when multiplied by itself. Take the number 36, for example. One of the factors of 36 is 6. If you multiply 6 by itself 6 6, you come up with 36, so 6 is the square root of 36. The number 36 has other factors, such as 18. But if you multiply 18 by itself 18x18, you get 324, not 36. So 18 isn't the square root of 36.

All whole numbers are grouped into one of two camps when it comes to roots:

» **Perfect squares:** Only a few whole numbers, called *perfect squares*, have exact square roots. For example, the square root of 25 is 5.
» **Irrational numbers:** Other whole numbers have square roots that are decimals that go on forever and have no pattern that repeats (nonrepeating, nonterminating decimals), so they're called *irrational numbers*. The square root of 30 is 5.4772255 with no end to the decimal places, so the square root of 30 is an irrational number.

Perfect Squares:

Square roots can be challenging to find without a calculator, but because you can't use a calculator during the test, you will have to use your mind and some guessing methods. To find the square root of a number without a calculator, make an educated guess and verify your results.

To use the educated-guess method (see the next section), you have to know the square roots of a few perfect squares. One good way to do this is to memorize the squares of the square roots 1 through 12:

» 1 is the square root of 1 (1x1=1)
» 2 is the square root of 4 (2x2=4)
» 3 is the square root of 9 (3x3 =9)
» 4 is the square root of 16 (4 x4 =16)
» 5 is the square root of 25 (5x 5= 25)
» 6 is the square root of 36 (6 x6 =36)
» 7 is the square root of 49 (7 x7= 49)
» 8 is the square root of 64 (8 x8 =64)
» 9 is the square root of 81 (9x 9= 81)
» 10 is the square root of 100 (10 x10 =100)
» 11 is the square root of 121 (11 x11 =121)
» 12 is the square root of 144 (12 x12= 144)

Tips for Taking Math Tests to Help You on Your Mathematical Journey:

As with the other ASVAB subtests, it may be wise to err on caution when taking the Mathematics Knowledge subtest, especially if you're taking the paper-and-pencil version. Choose any answer if you're doubtful of it and aren't in a hurry at the end of the subtest. You have no possibility of correctly answering if you don't. However, if you try for it, your chances rise to 25% or 1 in 4. The following sections contain some advice that will help you increase those chances even if the issue stumps you.

Knowing What the Question is Asking:

The Mathematics Knowledge subtest presents the questions as straightforward math problems, not word problems, so knowing what the question asks you to do is relatively easy. However, reading each question carefully and paying particular attention to plus (+) and minus (–) signs (which can change the answer) is still important. Finally, ensure you do all the calculations needed to answer correctly.

Check out this example:

Find the value of $\sqrt{81^2}$.
(A) 9
(B) 18
(C) 81
(D) 6,561

If you're in a hurry, you may put 9 down as an answer because you remember that the square root of 81 is 9. Or, in a rush, you may multiply 9 (the square root of 81) by 2 instead of squaring it, as the exponent indicates you should. Or you may multiply 81 by 81 to get 6,561 without remembering that you must find the square root, which gives you the correct answer, Choice(C). So make sure you perform all the operations needed (and that you perform the correct operations) to find the right answer. Here, noticing that you're both squaring 81 and taking the square root of 81^2 should make it easy to recognize that the answer is just 81 without calculating the multiplication.

Getting Clear on What You're Trying to Solve:

Read the question carefully right out of the gate. Some questions can seem out of your league at first glance, but if you look at them again, a light may go on in your brain. For example, suppose you get this question:

Solve for s: $s = \frac{2}{5} \times \frac{1}{2}$
(A) $2\frac{1}{2}$
(B) 2
(C) $\frac{1}{5}$
(D) $\frac{1}{10}$

At first glance, you may think, "Oh, no! Solve for unknowns. I don't remember how to do that!" But if you look at the question again, you may see that you're simply multiplying a fraction.

So you take ⅖ times ½ and arrive at 2/10, but you should reduce the fraction to get ⅕. So, the correct answer choice is (C).

Trying to Figure out What You Can and Guessing the Rest:

Sometimes, a problem requires multiple operations to arrive at the correct answer. If you don't know how to do all the operations, don't give up. You can still narrow down your choices by doing what you can.

Suppose this question confronts you:

What's the value of $(0.3)^3$?

(A) 0.0027
(B) 0.06
(C) 0.000027
(D) 0.0009

Let's say you have forgotten how to multiply decimals. There is hope yet! You must multiply 0 03 0 03 0 03 to recall how to utilize exponents. Therefore, your answer will contain a 27 if you simplify the problem by multiplying 333 without caring about those pesky zeros.

You can see from this nugget of knowledge that Option (B), which adds 0.03 to 0.03, is incorrect. Additionally, it implies that Option (D), which adds 0.03 and 0.03, is incorrect. With two options available, your likelihood of choosing the correct response has increased to 50%! Put the decimal points back in and multiply 333 to obtain 27. Move the decimal from 27 six places to the left to get 0.000027 since you need to make up six places. The right solution is option (C).

Using the Process of Elimination:

Another method when you run into questions and draw a total blank is to plug the possible answers into the equation and see which works. Say the following problem is staring you right in the eyes:

Solve for x: $x - 5 = 32$

(A) $x = 5$
(B) $x = 32$
(C) $x = -32$
(D) $x = 37$

If you're stumped and can't think of any possible way of approaching this problem, simply plugging in each of the four answers to see which one is correct is your best bet.

Choice (A): 5-5=32, which you know is wrong
Choice (B): 32-5=32, which is wrong
Choice (C): 32-5=32, which is wrong
Choice (D): 37-5=32, which is correct

Remember that entering all the solutions takes time, so keep this strategy for after you've finished answering all the questions on the ASVAB with pencil and paper. You cannot skip a question on the computer version of the exam, so plan your time carefully. If you're short on time, guess and carry on. The answer to the following query might be simple for you.

Questions for Practicing Math Skills and Operations:

Use these practice problems to exercise your math skills. You can check your answers and determine how much time you need to learn the fundamentals of mathematics before moving on to more difficult calculations. These questions are categorized as easy, medium, and hard. On the ASVAB, remember that you'll probably encounter a question of medium difficulty up front; the questions that come after depend on how well you perform on the first question.

ASVAB Mathematics Knowledge Practice Questions

Q1. What is the value of x in the equation $2x + 5 = 17$?
A) 6
B) 7
C) 8
D) 9

Q2. If a triangle has sides of lengths 5, 12, and 13, what type of triangle is it?
A) Equilateral
B) Isosceles
C) Right
D) Scalene

Q3. What is the area of a circle with a radius of 4 units? (Use π = 3.14)
A) 50.24 square units
B) 25.12 square units
C) 16 square units
D) 12.56 square units

Q4. If 5x − 3 = 2x + 7, what is the value of x?
A) 2
B) 3
C) 4
D) 5

Q5. A rectangle has a length of 8 cm and a width of 3 cm. What is its perimeter?
A) 22 cm
B) 24 cm
C) 26 cm
D) 28 cm

Q6. What is the slope of a line that passes through the points (2, 5) and (6, 9)?
A) 1
B) 2
C) 3
D) 4

Q7. If a square has an area of 64 square inches, what is the length of each side?
A) 8 inches
B) 16 inches
C) 32 inches
D) 64 inches

Q8. In the equation y = 3x + 4, what is the value of y when x = 2?
A) 6
B) 8
C) 10
D) 12

Q9. What is the volume of a cube with edges measuring 5 units each?
A) 25 cubic units
B) 75 cubic units
C) 125 cubic units
D) 150 cubic units

Q10. If the sum of the interior angles of a polygon is 1800 degrees, how many sides does the polygon have?
A) 10
B) 11
C) 12
D) 13

ASVAB Mathematics Knowledge Practice Questions - Set 2

Q1. What is the value of y in the equation 3y − 4 = 11?
A) 3
B) 4
C) 5
D) 6

Q2. A rectangle has an area of 56 square units and a length of 8 units. What is its width?
A) 4 units
B) 5 units
C) 6 units
D) 7 units

Q3. What is the slope of the line represented by the equation y = −2x + 3?
A) −3
B) −2
C) 2
D) 3

Q4. If the circumference of a circle is 31.4 units, what is its radius? (Use π ≈ 3.14)
A) 5 units
B) 10 units
C) 15 units
D) 20 units

Q5. In an isosceles triangle, if two sides are each 10 units long, what is the length of the base if the perimeter is 30 units?
A) 5 units
B) 10 units
C) 15 units
D) 20 units

Q6. What is the value of $8^2 - 3^3$?
A) 37
B) 43
C) 49
D) 61

Q7. If the sum of the angles in a quadrilateral is 360 degrees, what is the measure of an angle if the other three angles measure 90, 95, and 85 degrees?
A) 85 degrees
B) 90 degrees
C) 95 degrees
D) 100 degrees

Q8. What is the median of the set of numbers {3, 8, 12, 17, 22}?
A) 8
B) 12
C) 17
D) 22

Q9. If 3/4 of a number is 15, what is the whole number?
A) 18
B) 20
C) 22
D) 24

Q10. In a geometric sequence, if the first term is 2 and the common ratio is 3, what is the third term in the sequence?
A) 6
B) 12
C) 18
D) 24

ASVAB Mathematics Knowledge Practice Questions - Set 3

Q1. What is the sum of the interior angles of a pentagon?
A) 360 degrees
B) 540 degrees
C) 720 degrees
D) 1080 degrees

Q2. If $x^2 - 5x + 6 = 0$, what are the values of x?
A) 1 and 6
B) 2 and 3
C) 3 and 4
D) 4 and 5

Q3. What is the greatest common divisor (GCD) of 48 and 64?
A) 8
B) 12
C) 16
D) 32

Q4. A cylinder has a radius of 3 units and a height of 7 units. What is its volume? (Use $\pi \approx 3.14$)
A) 63 units^3
B) 197.82 units^3
C) 211.68 units^3
D) 317.46 units^3

Q5. What is the value of 2/3 of 27?
A) 9
B) 12
C) 18
D) 24

Q6. If a car travels 300 miles in 5 hours, what is its average speed in miles per hour?
A) 50 mph
B) 55 mph
C) 60 mph
D) 65 mph

Q7. What is the result of simplifying $3x^3 / 9x$?
A) $x^2 / 3$
B) x^2
C) $3x^2$
D) $9x^2$

Q8. In a right triangle, if one angle is 30 degrees, what is the measure of the other acute angle?
A) 30 degrees
B) 45 degrees
C) 60 degrees
D) 90 degrees

Q9. What is the next number in the sequence 2, 6, 18, 54, ...?
A) 108
B) 162
C) 216
D) 324

Q10. If a square's perimeter is 48 units, what is the length of one of its sides?
A) 8 units
B) 10 units
C) 12 units
D) 16 units

ASVAB Mathematics Knowledge Practice Questions - Set 4

Q1. If $4x + 3 = 19$, what is the value of x?
A) 2
B) 3
C) 4
D) 5

Q2. What is the area of a triangle with a base of 10 units and a height of 6 units?
A) 20 square units
B) 30 square units
C) 40 square units
D) 60 square units

Q3. Which of the following is a prime number?
A) 21
B) 29
C) 35
D) 49

Q4. In a 45-45-90 right triangle, if one leg measures 7 units, what is the length of the hypotenuse?
A) $7\sqrt{2}$ units
B) 10 units
C) 14 units
D) $14\sqrt{2}$ units

Q5. What is the simplified form of the expression 4x^4y^3 / 2xy?
A) 2x^3y^2
B) 2x^4y^3
C) 4x^3y^2
D) 8x^3y^2

Q6. If the exterior angle of a regular polygon is 30 degrees, how many sides does the polygon have?
A) 6
B) 8
C) 10
D) 12

Q7. What is the next number in the arithmetic sequence 3, 8, 13, 18...?
A) 21
B) 23
C) 25
D) 28

Q8. If 3y - 4 = 2y + 5, what is the value of y?
A) 1
B) 5
C) 9
D) 11

Q9. What is the product of the roots of the equation x^2 - 7x + 10 = 0?
A) -10
B) 7
C) 10
D) 17

Q10. A rectangular prism has dimensions 4 units by 3 units by 5 units. What is its volume?
A) 20 cubic units
B) 30 cubic units
C) 40 cubic units
D) 60 cubic units

ASVAB Mathematics Knowledge Answer Key

Q1. A) 6 - Solving 2x + 5 = 17 gives x = 6.

Q2. C) Right - Satisfies Pythagorean theorem: $(5^2 + 12^2 = 13^2)$.

Q3. A) 50.24 square units - Area of a circle: $\pi r^2 = 3.14 \times 4^2$.

Q4. B) 3 - Solving 5x − 3 = 2x + 7 gives x = 3.

Q5. A) 22 cm - Perimeter of a rectangle: 2(length + width) = 2(8 + 3).

Q6. A) 1 - Slope = (9 - 5) / (6 - 2) = 1.

Q7. A) 8 inches - Side of a square: √64 = 8 inches.

Q8. C) 10 - Substitute x = 2 into y = 3x + 4.

Q9. C) 125 cubic units - Volume of a cube: $side^3 = 5^3$.

Q10. D) 13 - Solving 180(n - 2) = 1800 gives n = 13.

ASVAB Mathematics Knowledge Answer Key - Set 2

Q1. C) 5 - Solve 3y - 4 = 11 to find y = 5.

Q2. D) 7 units - Area = length × width, so 56 = 8 × width, giving width = 7 units.

Q3. B) -2 - The slope of y = -2x + 3 is the coefficient of x, which is -2.

Q4. A) 5 units - Circumference = 2πr. With a circumference of 31.4, r = 31.4 / (2 × 3.14) = 5.

Q5. B) 10 units - In an isosceles triangle with a perimeter of 30, the base is 30 - 2 × 10 = 10 units.

Q6. C) 49 - Calculate $8^2 - 3^3$ = 64 - 27 = 49.

Q7. D) 100 degrees - Sum of angles in a quadrilateral is 360, so the fourth angle is 360 - (90 + 95 + 85) = 100 degrees.

Q8. B) 12 - The median of {3, 8, 12, 17, 22} is the middle number, 12.

Q9. B) 20 - If 3/4 of a number is 15, the number is 15 / (3/4) = 20.

Q10. C) 18 - Third term in the geometric sequence is $2 \times 3^2 = 18$.

ASVAB Mathematics Knowledge Answer Key - Set 3

Q1. B) 540 degrees - Sum of interior angles of a pentagon = 180° × (5 - 2) = 540°.

Q2. B) 2 and 3 - x^2 - 5x + 6 factors to (x - 2)(x - 3), so x = 2, 3.

Q3. C) 16 - The GCD of 48 and 64 is 16.

Q4. C) 211.68 units³ - Volume of a cylinder = $\pi r^2 h$ = 3.14 × 3^2 × 7 ≈ 211.68 units³.

Q5. C) 18 - 2/3 of 27 is 27 × (2/3) = 18.

Q6. C) 60 mph - Average speed = total distance / total time = 300 miles / 5 hours = 60 mph.

Q7. B) x^2 - Simplifying 3x^3 / 9x gives x^2.

Q8. C) 60 degrees - The other acute angle in a right triangle with one angle of 30° is 60°.

Q9. B) 162 - The sequence is a geometric progression with a common ratio of 3. Next number is 54 × 3 = 162.

Q10. C) 12 units - Perimeter of a square is 4 times the side length. So, side = 48 units / 4 = 12 units.

ASVAB Mathematics Knowledge Answer Key - Set 4

Q1. D) 5 - Solve 4x + 3 = 19 to find x = 5.

Q2. B) 30 square units - Area of a triangle = 1/2 × base × height = 1/2 × 10 × 6.

Q3. B) 29 - 29 is a prime number as it has only two distinct positive divisors: 1 and itself.

Q4. A) 7√2 units - In a 45-45-90 triangle, hypotenuse = leg × √2 = 7 × √2.

Q5. A) 2x^3y^2 - Simplify 4x^4y^3 / 2xy to get 2x^3y^2.

Q6. D) 12 - Number of sides of a polygon = 360° / exterior angle = 360° / 30° = 12.

Q7. B) 23 - Next number in the sequence 3, 8, 13, 18... is 18 + 5 = 23.

Q8. C) 9 - Solving 3y - 4 = 2y + 5 gives y = 9.

Q9. C) 10 - Product of the roots of x^2 - 7x + 10 = 0 is c/a = 10/1 = 10.

Q10. D) 60 cubic units - Volume of a rectangular prism = length × width × height = 4 × 3 × 5.

CHAPTER 10
Electronics Information (EI)

Electron Flow Theory:

Atomic knowledge is the foundation for a thorough grasp of electricity. Atoms are the smallest units into which an element may be divided while still retaining its characteristic characteristics, and they make up matter. Even smaller pieces, known as subatomic particles, are created when an atom is split down.

The nucleus contains two subatomic particles: protons and neutrons. Neutrons are neutrally charged, whereas protons have a positive charge. The majority of the mass of an atom is contained in its nucleus, which is also its heaviest component.

The third category of a subatomic particle is an electron, which revolves around the nucleus and has a negative charge. One electron exists for every proton in the nucleus of a neutral atom. All electricity is the movement of electrons, including static electricity and lightning strikes.

Around the nucleus, the electrons inhabit different energy levels known as shells; as one fills up, a new one starts. The number of electrons in an atom's valence shell—its outermost protective layer—determines whether an element is a conductor, a semiconductor, or an insulator. The valence shell of many elements can only accommodate a maximum of eight electrons.

Boron, a Conductor; Silicon, a Semiconductor; and Phosphorus, an Insulator

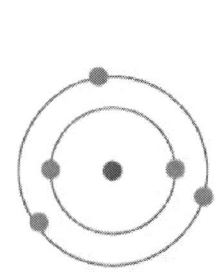

Boron: 5p 6n 5e
3 valence electrons

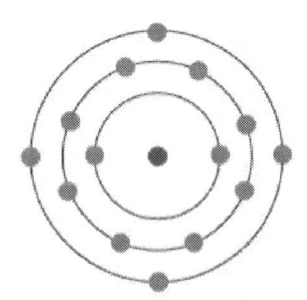

Silicon: 14p 14n 14e
4 valence electrons

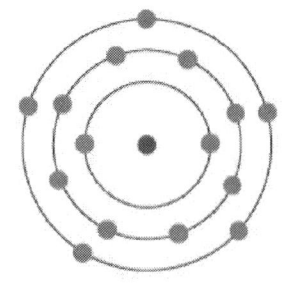

Phosphorus: 15p 15n 15e
5 valence electrons

A conductor is a substance that enables unrestricted electron flow. Since each valence shell has more open spaces than electrons, all conductors share one or more mobile valence electrons, free to migrate from one atom to another. An insulator has a valence shell that is fully or nearly filled. Because all its electrons are firmly bound and will not escape their crammed valence shells, it barely conducts any electricity.

Despite not being a good conductor or insulator and having a valence shell that is exactly half full, semiconductors have certain unique features that make them excellent materials for creating electronic components. Imagine the electrons as drops of water running through a conduit to understand how electricity functions easily. Pressure being applied at one end causes the water to flow in the other direction, causing this flow.

Similarly, when a conductor is connected across the battery's terminals, its negative terminal repels electrons and forces them toward the positive terminal. (A fundamental principle of electricity states that like charges repel one another while opposite charges attract.) Electric current is the resultant flow of electrons via a conductor.

Current:

Current is the measure of how quickly electrons move across a conductor. The passage of electricity has a current just like the surge of water through a conduit. When a faucet is turned on, the diverse volumes of water released simultaneously make it simple to discern between a trickle and a thundering flood.

Similarly, the amount of charge flowing through per unit of time is used to quantify electrical current. An ampere is one coulomb (C), the fundamental unit of electrical charge, flowing past a specific point in one second. Current is measured in amperes, or amps for short. A is the standard symbol for amperes. Five amps (5 A) of electricity would flow through a place every second if five coulombs of charge were to pass through it.

Voltages:

For electrons to move through a circuit, there needs to be a force that moves them. In a line, water will move from a high-pressure area to a low-pressure area, but nothing will move if the pressure is the same at both ends. The same goes for electricity. For electrons to move through a conductor, there must be "pressure" from electricity on the wire.

Electromagnetic repulsion from a big negative charge, like that at the negative end of a battery, causes "pressure" for an electron. This is because there are a lot of electrons at this negative end, and they are all pushing against each other. When a piece of conducting wire connects a battery's positive and negative terminals, this is called "completing the circuit." Electrons are pushed from the negative terminal into the wire, along the conducting electrons already there, and so on down the line.

Eventually, every electron will reach the positive terminal, with few electrons, and many attractive protons pull the electrons in. This will continue until all the extra electrons at the negative end are used up and the battery dies.

Voltage is the strength of electricity, and it is measured in volts, which is written as a capital letter V. When a higher voltage is given to a conductor, the electrical pressure on the electrons will be higher, and the rate at which electrons flow through the conductor will tend to be faster. So, there is a straight relationship between voltage and current. To be more precise, voltage is the difference in electric pressure between two places.

Electrons move from areas with higher electric potential (the correct term for this pressure-like amount) to areas with lower electric potential (lower pressure). Because of this, voltage is also called the difference in the electric potential. It is also sometimes called the electromotive force. An expert would use a voltmeter to measure voltage.

Resistance:

Resistance is anything that stops the flow of electricity. Resistance is measured in ohms. One ohm is the resistance that lets one ampere of current run through a conductor when one volt of electrical pressure is applied. The symbol for an ohm is the Greek letter omega.

An ohmmeter is used to measure resistance. When resistance goes up in a medium, the current flow goes down. This means that the relationship between current and resistance is backward. The better a material is as a conductor, the less resistance it has. But remember that voltage and current are directly linked. Whenever the voltage goes up, the current will also go up.

So, a higher voltage must be used to get the same amount of current from materials with high resistance or that don't carry electricity well. Some conductors have more resistance than others, but all have some. Silver, copper, and aluminum are the three most commonly used conductors.

Silver is the best conductor because it has the least amount of resistance. Silver is expensive, so copper is the next best pick. Copper is much cheaper than silver but has a slightly higher barrier than silver. Copper is used to make most electricity wires and cables right now. Aluminum has a higher resistance than copper and some other qualities that make it even less desirable for electrical use. Aluminum is used a lot in home wiring but only in a few specific places.

EXCEPTIONS AND ADDITIONAL NOTES:

The comparison of electron flow to water moving through a pipe is helpful, but like all comparisons, it breaks down in some places. When a pipe breaks, water sprays out, but if a wire in an electronic device breaks, the flow of electricity stops along that line. That's because energy moves through the metal of a wire, while water moves through a hollow pipe. Without something to move through, electrons can't move forward, and air molecules don't move electrons very well.

Electricity won't flow through the air unless the gap is very small (which is how spark plugs are made) or the voltage is very high (like when lightning strikes). Conventional current is another interesting quirk in the study of electricity. It is defined by the (imaginary) flow of positive charge and goes opposite to real electron flow. This strange idea arose because seeing electricity at a tiny level is hard. On the other hand, it is easy to tell which way it is going because you can see and feel it.

Electricity, on the other hand, is a different story. When scientists started seriously studying electricity, they knew little about atoms. Then, they realized that the movement of charges in electricity could be caused by either a flow of negative electrons in one direction or positive protons in the other.

Later, they discovered that protons stay in the center and don't move, but this was already how the current direction was set. So, if a circuit has a normal current that moves in a clockwise direction, electrons are going in a counterclockwise direction around the circuit.

Today, we can talk about either regular current or electron flow, but when we switch from one to the other, we must remember to change the way. It's easy to think of electrical current as a flow of electrons, but one electron doesn't always go around a circuit. Often, this is more like a shuttle race, as the charge is passed from one electron to the next. So, some sources might talk about the flow of a charge.

Circuits:

For electric current to move, it needs a path to follow or a circuit. A circuit is a path or loop that energy can go through. An electrical circuit needs a voltage source, something to power, and wires to connect the load to the voltage source. When these three things are joined so that electricity can flow, the circuit is closed.

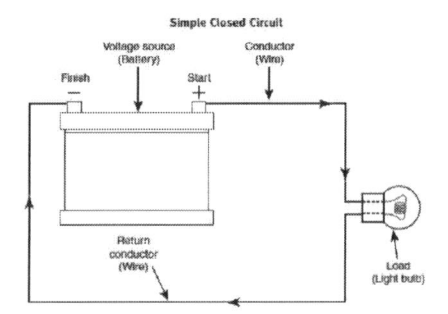

A load is a source of resistance that changes electrical energy into another form of energy. A load is something like a light bulb. It has resistance and changes the energy from electricity into light and heat. Electric motors, heated elements, and solenoids are all other types of loads. If the resistance of a load went down, the current moving through the circuit would go up. But no current would pass if the circuit was broken like a wire (called a "conductor") being cut.

This is known as an "open circuit." When current is flowing, each electron carries the voltage along the length of the conductor, forcing the electrons ahead of it to keep moving forward while being pushed themselves by the electrons behind it. When the line stops, it stops everywhere. When there is no clear way for electrons to reach their target, like when there is a gap in an open circuit, electron flow stops, and there is no current.

OHM'S LAW:

Ohm's law describes the link between a wire's voltage, current, and resistance. Ohm's law says that the voltage (potential difference) in volts equals the current in amperes times the resistance in ohms.

The formula represents this:

$V = IR$

V stands for "voltage" or "potential difference." I stands for "current," and R stands for "resistance." If the I symbol looks strange, consider current as the strength of charge flow. Current is the amount of charge moving past a point per unit of time. It's easier to remember the other two. The power source is where the circuit starts and where it ends.

Electrons must move through the circuit because the negative side makes them want to go away. On the other hand, the positive side helps by attracting electrons, and an electron that leaves the negative side will finally travel through the whole circuit before getting to the positive side.

Remember that voltage is just electric pressure on a charged object, so the voltage source is the one that makes the electrons move in a circuit. In other words, Ohm's rule shows that a non-zero voltage will produce a certain amount of current in a closed circuit with a given resistance.

Voltage sources are things like batteries, which have a positive and a negative end—for example, a 12 V battery. An electricity outlet in a home also provides voltage from the power plant. They are also called power sources because the electricity they make has energy.

The conductor is the wire that links the power source to the load(s). To keep things simple, the line is often thought to have no resistance. Even though this isn't true, the resistance in a conducting wire is usually small enough compared to the load that can be ignored for most estimates.

This is because Ohm's law says that when the resistance of a conducting line is zero, the voltage across it must also be zero. This means the electrical pressure does not change from one side to the other.

Remember that a load is any part with resistance and changes electrical energy into another form. Ohm's law says that if the resistance of a load goes down, the current moving through the circuit will go up. But no current would pass if the circuit was broken like a wire (called a "conductor") being cut. This is known as an "open circuit." Ohm's law can also explain this since the break or gap in the circuit is like a load with infinite resistance (remember from the last part that air is a terrible conductor).

SERIES CIRCUITS:

A series circuit is an electrical circuit with only one way for current to move. Therefore, any break (gap) in the circuit will stop the flow of electricity through the whole circuit. A series circuit is the simplest circuit, with one voltage source and one load linked by wires.

Series Circuit

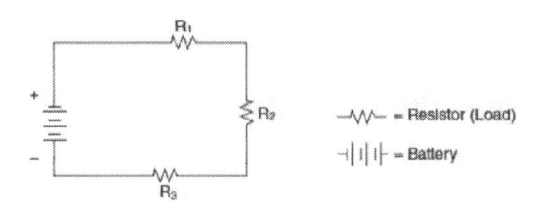

—⋀⋀— = Resistor (Load)
⊣|ı|ı|⊢ = Battery

In a series circuit, there is only one way for the current to move so that the current flow will be the same everywhere. For example, it wouldn't make sense if one million electrons left the negative end of a battery every second, but two million electrons came in at the positive end. It would be like a plane leaving New York and arriving in Miami with twice as many people on board. So, the current going through any single load is the same in a series circuit.

$I_1 = I_2 = I_3 = ...$

On the other hand, what would happen if a series circuit had more than one load? When two or more loads are added to a series circuit, their resistances add up to make the total or effective resistance across the circuit.

$R_{tot} = R_1 + R_2 + R_3 ....$

Voltage is the change in electric potential (also called "electric pressure") between two points, such as between a voltage source's negative and positive terminals. It can also be measured across a single load. This is also called the voltage drop across a load, and the sum of the voltage drops across each load in a series circuit is equal to the total voltage of the whole circuit, which is the same as the voltage of the voltage source.

$$V_{tot} = V_1 + V_2 + V_3 \ldots$$

PARALLEL CIRCUITS:

More often than not, loads are connected in parallel. With this setup, each load is placed on its path. Even if one of these paths had a break or gap, the current would still move through the other paths, so the circuit would still be closed.

Parallel Circuit

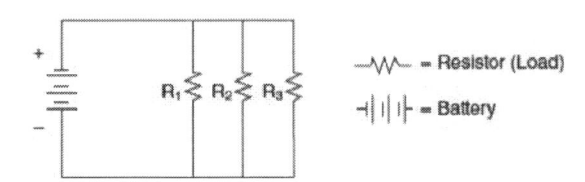

Parallel circuits are the exact opposite of series circuits. Each parallel branch of a parallel circuit has the same voltage, but the current flows differently. Since voltage is a measure of the difference in electric potential between two places, the voltage between two points in a circuit connected by two paths must also be the same.

$$V_1 = V_2 = V_3 = \ldots$$

On the other hand, when parallel routes exist, the current flow is divided. Therefore, the statement that electricity takes the route of least resistance is frequently incorrect. In reality, this is untrue.

In reality, electricity travels down every conceivable path, but a greater percentage of electrons will choose a low-resistance route than a higher-resistance one. The sum of the currents flowing through each closed path determines the overall current flowing through a parallel circuit.

$$I_{tot} = I_1 + I_2 + I_3 \ldots$$

What happens to a circuit's effective resistance when many loads are connected in parallel? It's human nature to assume that increasing the load will automatically increase resistance. However, this is not the case. At first glance, this seems a little less logical, so think of an analogy. A well-known retailer is now having significant sales.

Customers jam one of the aisles to score the greatest offer. Moving forward is tough, and moving slowly through the aisle is slowed down by the shoving shoppers, abandoned shopping carts, and containers of goods waiting to be shelved.

In this example, the individuals are represented by electrons, and their movement rate represents the current value. Unexpectedly, a store employee removes a stack of goods obstructing an adjacent aisle. Unfortunately, there is a ton of clutter blocking this aisle as well.

Since it contains more items, it might be more resistant than the original aisle. However, some customers shop in that narrow but less busy lane. With the construction of that parallel corridor, more customers are entering and exiting the business at any given time.

Similar to adding a load in series, adding a load in parallel reduces a circuit's overall or effective resistance because, despite having resistance, the additional path it creates does not affect the existing one. It's a little trickier, but the parallel resistance addition formula is helpful.

$$1/R_{tot} = 1/R_1 + 1/R_2 + 1R_3$$

SERIES-PARALLEL CIRCUITS:

The series-parallel circuit is the most common configuration. An on/off switch is one example of a component wired in series with a number of loads connected in parallel in a series-parallel circuit.

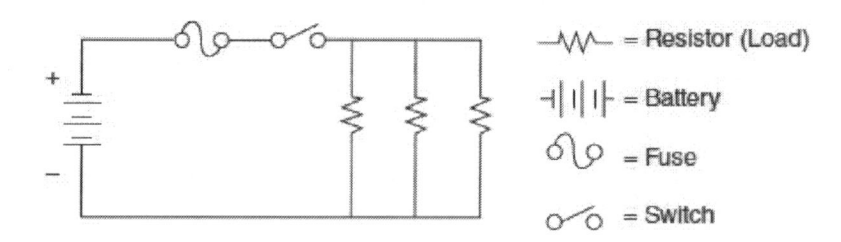

Circuits used in household wiring are typically series-parallel. For example, wall outlets are wired in series but fed by a circuit breaker, represented in the diagram as a fuse. The outlets' electricity will be turned off if the circuit breaker is switched off. However, whether or not there are loads plugged into the outlets, voltage is given to all of them if the circuit breaker is on.

A series-parallel circuit's total effective resistance can be calculated step by step. To reduce the number of predicted effective resistances in the circuit to a single value, apply the appropriate formula to each set of connected resistors in series or parallel.

ELECTRICAL POWER:

An electric circuit's real energy supply and consumption rate is referred to as its electrical power. Calculating power, which is measured in watts (joules per second), involves multiplying the voltage (in volts) applied to the circuit by the current flow rate (in amperes).

The formula represents this:

$P = IV$

Each electron carries energy as it moves across a wire or circuit. The amount of energy is determined by the voltage used. The energy of electrons traveling through a high voltage is greater.

On the other hand, more electrons are going by each instant when the current is larger.

A smaller number of electrons per moment (lower current) with a higher amount of energy per electron (higher (D) 12 A voltage) or a larger number of electrons per moment (higher current) with a lower amount of energy per electron (lower voltage) can both deliver energy at an equivalent rate.

This formula applies to the rate of energy production (power generated by a generating plant), the rate of energy consumption (power used by an appliance or load), as well as the rate of energy delivery (power found in long-distance power lines).

The term "consumption" is a little deceptive because energy never runs out; it simply shifts forms. The actual energy consumption of a 40-watt (W) light bulb is less than 40 joules per second. Every second, it converts 40 joules of electrical energy into 40 joules of light and heat energy.

STANDARD ELECTRICAL UNITS AND THE METRIC SYSTEM:

It's crucial to remember that the amounts used in Ohm's law and other electrical formulas must be represented in ohms, amperes, volts, and watts. Therefore, before calculating any of these amounts, they should be converted if supplied in other units.

Metric prefixes can be added to any of these base units. Megawatts (MW) are a common unit of measurement for smaller power plants, whereas cardiologists use milliamps (mA) to gauge the tiny electrical signals produced by the human heart.

Structure of Electrical and Electronic Systems:

You have now gained a rudimentary understanding of an electric circuit's operational and mathematical principles. You will advance these talents in the subsequent part. The explanation of various current kinds, specific safety tips, and components that give circuits unique features and allow them to carry out particular jobs are all included in this section.

AC VS. DC:

The current comes in two flavors. Direct current (DC) and alternating current (AC) are what they are. DC, or direct current, only flows in one direction through a conductor. This is the kind of current that a battery produces. Many electrical devices, including computers and cell phones, use DC power. As it passes through a conductor, AC is a current that frequently reverses direction (moves back and forth). Alternating current is delivered through household electrical sockets.

The delivery of electricity to your home or place of business from the power plants is more effective with AC. In North America, the frequency of AC is always 60 cycles per second (Hz). This is because voltage would begin at zero, rise to a maximum level in one direction, drop to zero, rise to a maximum level in the opposite direction, and then rise to zero once more to complete one cycle. The signal's frequency is expressed in hertz (Hz), which measures how many full cycles occur in a second.

GROUND:

Residential wiring and electrical equipment should always be grounded. The area of a circuit with the lowest potential is represented by ground. Any "stray" electricity will travel along this path since resistance is minimal here, and the potential difference (voltage) between any point in the circuit and the ground point is the biggest.

This is crucial to prevent shock from internal circuit failure when conducting wires are degraded and external impacts like lightning.

The ground connection is a universal link in domestic wiring that guards against electrical shock. In the event of an internal short circuit, electrical current is directed away from panels and equipment by an earth ground (such as a copper rod pushed into the ground or buried conduit) connected to all of the wiring grounds.

IMPORTANT ELECTRIC AND ELECTRONIC COMPONENTS:

The properties of a particular electrical circuit will change depending on the types of components present. In this section, several of the major ones are introduced and discussed.

Resistors:

An electrical circuit's components don't all require the same voltage. It can occasionally be helpful to increase or decrease the current flowing through a circuit to control specific functions, such as a stereo's volume. A resistor is a part that can be used to restrict current and/or voltage.

As its name implies, a certain amount of resistance is created when current flows through a resistor, causing a voltage drop. As you already know, Ohm's law explains how resistance can affect current. Therefore, an increase in R will result in a decrease in current.

Resistor Symbols

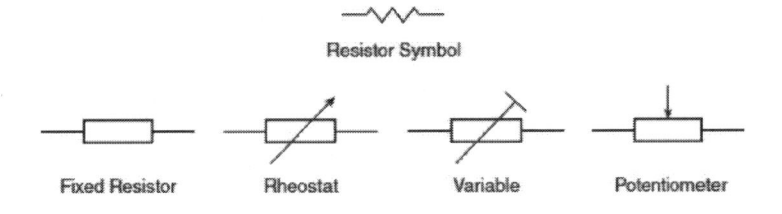

Resistor Symbol

Fixed Resistor   Rheostat   Variable   Potentiometer

The fixed resistor and the variable resistor are the two main types of resistors now in use. As their name suggests, fixed resistors have a fixed resistance. Resistances in variable resistors (rheostats or potentiometers) can be adjusted. Potentiometers help adjust the voltage drop across a circuit component. They can be used in electronic applications to adjust the TV, radio, or stereo volume. Rheostats can be used as light dimmers or to regulate the speed of small motorized devices.

They help regulate the current in a circuit. Fixed resistors allow us to calculate their resistance in ohms using a color band approach. The first color reveals the first number. The second color indicates the second number; the third indicates the multiplier or the number of 0s after the numbers, and so forth.

Fuses and Circuit Breakers:

The electric ground is crucial to a circuit's safety. Giving electricity another route away from the gadget helps reduce shock. Circuit breakers and fuses are additional critical safety components. Wires can overheat and catch fire if the current in a circuit grows too much.

A short circuit results when a conductor intentionally bypasses a load for some reason. The current flow will rise since the relationship between current and resistance is inverse. In addition, the insulation on the wires going to and from the load could get frayed, allowing the wires to come into contact with each other directly, which could be the root of the problem.

The only thing that happens if a short circuit happens in a section of a series circuit is that the load is bypassed because the short circuit is removed. But suppose the short circuit bypasses all the loads in series circuits or occurs at any load in parallel circuits. In that case, the resistance of the circuit as a whole is decreased to almost zero, and a potentially harmful current surge may happen. Fuse or circuit breaker protection for circuits is an excellent safety measure.

Fuses are thin wires that melt when the current reaches a set limit, stopping the energy flow. This shields the electric gadget from any possible harm. One drawback of employing a fuse is that it must be changed once it has melted or "blown" for the circuit to resume functioning. Fuses with various current ratings can be utilized depending on the device's requirements.

Fuses serve the same purpose. However, circuit breakers have the advantage of being reusable several times. They cost more to install and react to increases in current more slowly than fuses do. Although there are several kinds of circuit breakers, they all work on the same basic concept. When excessive current runs through a circuit, a bimetallic strip used in one circuit breaker bends away from its contact.

Breaking the circuit stops more current from flowing. A breach in the circuit is created by an electromagnet in a different type of circuit breaker. When the current reaches a specific level, the ferrous material becomes magnetic enough to open the circuit and stop the flow of electricity.

Capacitors:

Electrical storage devices are capacitors, also referred to as condensers. They are created by sandwiching a very thin insulator, or dielectric, between two metal conducting plates. Additionally, air can act as a dielectric.

Because the DC source generates an excess of electrons on the negative plate and a deficiency of electrons on the positive plate, a capacitor can store an electrical charge. Even when the voltage source is withdrawn, the charge in the capacitor is maintained by the electrostatic attraction between the positive and negative charges.

If a conductor is connected across the capacitor, a passage is made for electrons to flow from the negative plate to the positive plate, causing the capacitor to discharge itself.

A capacitor will permit AC but prevent DC from passing through it. This is why "charging up" a capacitor works best with DC. Capacitive reactance, expressed in ohms, is a capacitor's resistance to current flow. The frequency of the AC signal is inversely correlated with capacitive reactance. In other words, there is less resistance to the flow of AC across the capacitor at higher frequencies.

The farad serves as the unit of measurement for capacitance, which is the capacitor's capacity to store charge. Capacitance is denoted by the letter C. A farad has enough capacitance to store one coulomb of electrons at one volt of electrical potential.

SEMICONDUCTORS:

An element with four electrons in its valence shell is called a semiconductor. These elements are neither good conductors nor insulators because of the strong links between these four electrons and the nucleus. Silicon and germanium are two substances that are frequently recognized as semiconductors. Silicon and germanium are not highly helpful in their pure forms. However, a whole new universe of opportunities arises when impurities are added to its crystalline structure through doping. Pure silicon has a crystalline structure that is remarkably stable.

There are no free electrons to permit current flow because the four valence electrons in each silicon atom bond with all the valence electrons in the atoms nearby. This can be altered by "doping" phosphorus, arsenic, or antimony into the silicon's crystal structure. Since each of these elements has five electrons in its valence shell, it will connect to the other silicon atoms, leaving one free electron to move around the crystal. As a result, the silicon crystal becomes an N-type substance.

This new substance is electrically neutral and can conduct electricity despite free electrons. When silicon is doped with substances like boron or indium, which each have three electrons in their valence shells, the crystalline structure of silicon is left with a "hole" where an electron would typically be. Repeating regions with a net positive charge result from this. The formerly occupied places are left with "holes" as the free electrons enter these areas. Due to this doping, silicon is now referred to as a P-type material.

Diodes:

P-type and N-type materials are useless, but a diode is created when combined. A diode electrical one-way valve is created when P-type and N-type materials interact at their junction. In one way, the current can flow easily, but in the other, it is obstructed. Because current can only flow in one way, the battery's position within a circuit is crucial. When current can flow, it is said to be forward-biased.

The P-type material, known as the anode, should be linked to the battery's positive terminal. The N-type cathode material should be attached to the battery's negative terminal to make the circuit forward-biased.

Free electrons in the N-type material are repelled by the extra electrons at the battery's negative terminal and are drawn toward the junction. Current passes through the diode as the N-type material's electrons cross the junction and fill the P-type material's "holes." A different set of circumstances arises when the diode's connection to the battery is reversed.

Free electrons in N-type materials are attracted to the opposite charge on the battery's positive terminal, which causes them to travel away from the junction. As a result, current flow ceases since there is no longer any electron transfer at the junction due to the electrons having left. Therefore, the diode is currently biased in the opposite direction.

Diodes are used most frequently for rectification. AC to DC conversion is referred to as rectification. A diode will only pass the upper or lower half of the AC waves since it only permits current to flow in one direction. Half-wave rectification is the term for this.

Most DC power supplies are built on full-wave rectifiers made by connecting four diodes in a diamond pattern. Diodes are also employed for rectification in automobile alternators, where generated AC is transformed into DC and utilized in the vehicle's electrical system.

Transistor:

The transistor is an electrical component that has completely changed how computers, calculators, and radios are built. A transistor is exceptionally adaptable because it can function as an electrical switch, an amplifier, or a current regulator. A transistor is a solid-state device since it has no moving components and is constructed of solid silicon, which makes it incredibly dependable.

A single integrated circuit can include thousands of transistors, making it small. An integrated circuit comprises a single silicon chip that houses all required circuit components. The core of the logic operations that a computer performs to process data is transistors.

Transistors come in two varieties: NPN transistors and PNP transistors. An NPN transistor consists of two pieces of N-type material and a thin P-type material.

The opposite is a PNP transistor, which consists of two P-type parts with an N-type piece sandwiched in between. Wires are connected to the three elements of semi-conductor material inside the transistor. Every item has a name. The base is always the central component, and the collector and emitter are the two outside components.

The emitter is indicated by an arrow on the transistor's symbol. The arrow's direction can determine what kind of transistor it is. The fact that the arrow always points in the general direction of the N-type material makes it easy to recall which symbol designates which sort of transistor.

Symbols for NPN and PNP Transistors

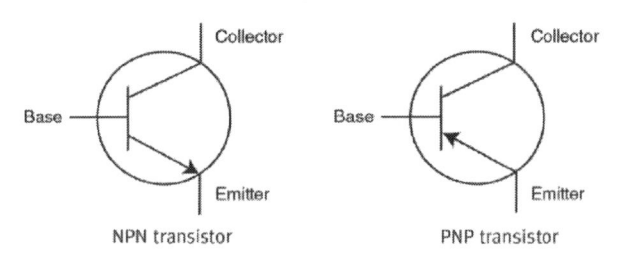

NPN transistor / PNP transistor

The transistor uses a tiny quantity of current to regulate much current. An NPN transistor can be "turned on" by applying a positive voltage to the base, which will cause a sizable current to flow from the collector to the emitter. The transistor turns off and stops the current flow when the positive voltage is removed from the base.

Current Flow in an NPN Transistor

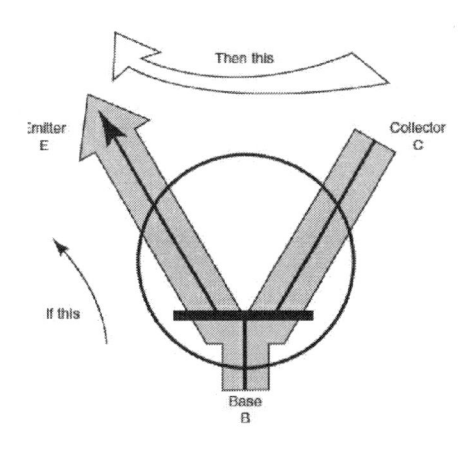

PNP transistors work opposite to NPN transistors. A PNP transistor requires a negative voltage at the base to turn it on, and the current then flows from the emitter to the collector.

Electricity and Magnetism:

Magnetism and electricity are intimately related. Magnetic fields are created by moving charges; moving electrons in a circuit is no different. The interaction between electricity and magnetism has proven helpful in various applications, including circuit breakers, transformers, motors, generators, microphones, and doorbells.

A wire that carries current produces its magnetic field.

The amount of recent passing through the wires determines how strong the field is. An electromagnet is created when the iron core of a ferrous material is magnetized by the magnetic field created when the wires are wrapped around it, and a current is run through them. The fundamental idea behind many electrical devices discussed above is this generated magnetism.

INDUCTORS:

A current-carrying wire can be wound into a coil to increase the strength of the magnetic field it creates. The magnetic field surrounding the coil gets much more vital if wound onto an iron core because iron conducts magnetic lines of force more effectively than air does.

The magnetic field develops rather slowly when current first passes through a coil. This is due to the counter-emf, or voltage generated by the growing magnetic field within the coil, which opposes the initial current flow.

The magnetic field shrinks when the current is turned off, and this decreasing magnetic field produces a voltage in the coil that maintains the current's flow. Electrical parts known as inductors display a characteristic known as self-induction, which is resistance to changes in current flow in a circuit.

Inductors thwart modifications in current flow. By producing a voltage that moves against the applied current, the inductor fights against the growing current. If recent drops occur, the inductor uses the coil's magnetic energy to counteract the drop and maintain current flow.

Induction is measured in a unit known as **henries**, and the symbol used to represent induction is L. Inductors work precisely opposite to capacitors in the sense that they allow DC to pass quickly but resist the flow of AC. This resistance to current flow is called inductive reactance and is measured in ohms. It increases with the increasing frequency of the AC signal.

TRANSFORMERS:

A transformer is employed in a circuit to change the voltage. To do this, a transformer makes use of an inductor's characteristics. An iron core is magnetized by alternating current running via wires encircling it, which creates a fluctuating magnetic field inside the core.

A nearby wire coil experiences a voltage due to this shifting magnetic field. A lesser or bigger voltage can be induced in the secondary coil depending on the main coil vs. the secondary coil's wire turns and the distance between the coils.

The coil that is attached to the source is the primary coil. The coil in which an electric current is induced is the secondary coil. More secondary coils indicate a higher voltage. The efficiency of the transformer in creating a voltage in the secondary coils increases with the distance between the secondary and main coils.

Transformers are particularly helpful for transferring electricity from power stations to buildings and commercial establishments. Transmission of low current, high voltage electricity is more energy efficient. However, when it gets to houses, it needs to be "stepped down" or reduced to the typical 120–240 volts that most appliances in our homes utilize.

BASIC ELECTRICAL MOTORS AND GENERATORS:

A permanent magnetic field will exert a force (push or pull) on a wire carrying current. It is possible to use this force to create usable mechanical energy. This is how a straightforward motor operates. Mechanical energy can be utilized to lift large objects, move the tires on a car, or spin a fan, to mention a few applications.

The quantity of current flowing, the wire's length, and the magnetic field's strength directly affect how much force is generated on the current-carrying wire. By increasing any of these, the force acting on the wire will rise, increasing its ability to perform work.

Of course, today's motors are more complicated than this, and many use electromagnets instead of just a wire to transfer the current, but the basic idea is still the same.

On the other hand, a generator is just a motor running in reverse. If a wire is connected in a complete circuit, moving it in a persistent magnetic field causes it to conduct current. As a result, a generator transforms mechanical energy into electrical energy.

Electronics Information Practice Set 1:

1. How much current will flow in a circuit that has 60 mV applied to a 15 KΩ resistance?

(A) 0.004 mA
(B) 0.9 A
(C) 4.0 A
(D) 900 A

2. What voltage is required for 30 A to flow through a 60 kΩ resistance?

(A) 1,800 V
(B) 1.8 kV
(C) 18 kV
(D) 1.8 MV

3. What is the current in the 20-ohm resistor?

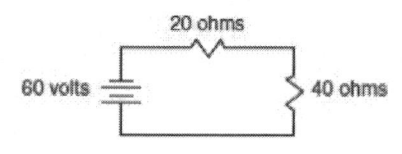

(A) ½ amp
(B) 1 amp
(C) 30 amps
(D) 100 amps

4. Given the image below, which is the valence shell in a copper Atom?

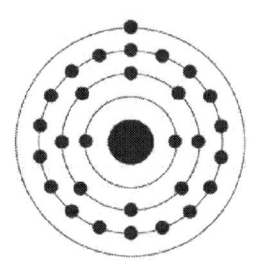

Copper atom (29 electrons)

(A) the first
(B) the second
(C) the third
(D) the fourth

5. Which is the collector in the transistor symbol below?

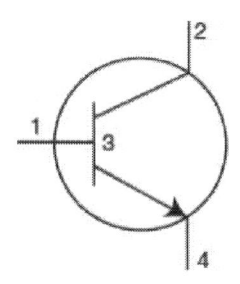

(A) 1
(B) 2
(C) 3
(D) 4

6. Current can be expressed in:
(A) kilohms
(B) kilovolts
(C) milliamperes
(D) millivolts

7. A circuit with a flowing current in it can be assumed to be
(A) open
(B) closed
(C) short
(D) dead

8. Which of the following symbols represents a polarized capacitor?

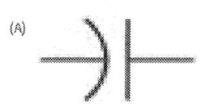

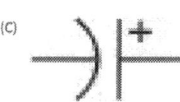

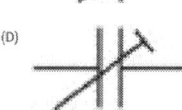

9. What is created When a P-type material is joined with an N-type material?
(A) a resistor
(B) a diode
(C) an inductor
(D) a capacitor

10. Which switch, if flipped, would not close the circuit in the diagram Shown?

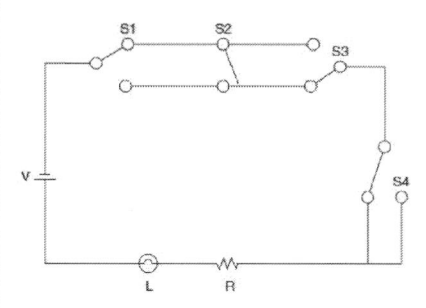

(A) S1
(B) S2
(C) S3
(D) S4

11. Given a circuit with a single resistor, adding two more resistors to the circuit, both with the same resistance value as the original resistor, one in parallel with the original resistor and one in series with them both, will cause the overall resistance to
(A) definitely increase
(B) definitely decrease
(C) stay the same
(D) either increase or decrease depending on the wiring order

12. Which of the following would not increase the rate of direct current flow, I, through a conductor?
(A) decreasing effective resistance via a variable resistor
(B) increasing voltage
(C) adding an inductor to the circuit (in series)
(D) adding a resistor to the circuit (in parallel)

13. A circuit includes three 2 Ω resistors wired in parallel. What is their effective resistance?
(A) ½ Ω
(B) ⅔ Ω
(C) Ω
(D) 6 Ω

14. Why is copper favored over other conducting materials like aluminum and silver?
(A) Copper has the lowest resistance.
(B) Copper has the highest resistance.
(C) Copper is inexpensive.
(D) Copper has low resistance relative to its cost.

15. A 6 V battery provides power to a flashlight whose 10 W light bulb is the only load in the circuit. What is the amount of current flow through the flashlight?
(A) 0.6 A
(B) 1.67 A
(C) 6 A
(D) 60 A

ELECTRONICS INFORMATION PRACTICE SET 2:

16. A 2.5 V battery is used to power a cordless phone. An ammeter shows the current flow is 30 A through the phone when it is on. How much power does the phone use?
(A) 0.067 W
(B) 12 W
(C) 32.5 W
(D) 75 W

17. A shop vac is plugged into a 120 V generator. Given a total current flow of 15 A, what must the total resistance of the vacuum cleaner be?
(A) 8 Ω
(B) 15 Ω
(C) 120 Ω
(D) 1800 Ω

18. Two identical resistors are wired in parallel in a circuit. The voltage drop across these resistors is 18 volts, and the total current in the circuit is 6 amps. What is the resistance of either resistor?
(A) 1.5 Ω
(B) 3 Ω
(C) 6 Ω
(D) 9 Ω

19. Which of the following is one of the reasons neighborhoods are connected to the power grid via electrical substations instead of directly wired into long-distance power lines?
(A) The substations use special components to connect aluminum power lines to copper household wiring.
(B) The substation transforms direct current to alternating current for household use.
(C) The substation transforms alternating current to direct current for household use.
(D) The substation transforms the efficient high-voltage power from the lines to a safer, lower value for household use.

20. A fixed resistor has a black, orange, and orange color band pattern.
What is its resistance?
(A) 33 ohms
(B) 3000 ohms
(C) 30,000 ohms
(D) 33,000 ohms

21. Given the following DC circuit, which has been open for some time, what will happen after the switch is closed?

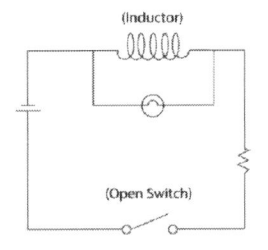

(A) The total current flow will be smaller when the lamp first lights, then increase dramatically.
(B) The lamp will light up brightly and then burn out.
(C) The countercurrent of the inductor will spontaneously cause the closed switch to open again.
(D) The power draw on the battery will be higher at first, then lower as the resistance decreases.

22. A European traveler brings his hair dryer to the United States. The specifications on the small appliance say it draws 10 amps when on a standard household circuit, which is 220 volts in Europe. What will it draw if connected directly to a U.S. power source of 110 volts?
(A) 5 amps
(B) 10 amps
(C) 15 amps
(D) 20 amps

23. A portable air compressor converts electric energy to the potential energy of pressurized air. A mechanics student calculates the total energy of the pressurized air at about 800,000 J. If it took 400 seconds to pressurize the machine fully, what must be the electric power rating of the compressor?
(A) 400 W

(B) 2000 W
(C) 800,000 W
(D) 320,000,000 W

24. Given a lamp wired in a simple AC household circuit, which of the following will cause the light to dim the most?
(A) A capacitor wired in parallel with the lamp
(B) An inductor wired in parallel with the lamp
(C) A capacitor wired in series with the lamp
(D) An inductor wired in series with the lamp

25. In the circuit diagram pictured, where the two lamps have identical loads, what will be the effect of closing the switch?

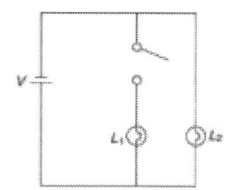

(A) L1 will light up, and L2 will shut off; the power draw on the battery will be the same.
(B) L1 will light up, and L2 will dim; the power draw on the battery will double.
(C) L1 will light up, and L2 will remain at the same brightness; the power draw on the battery will stay the same.
(D) L1 will light up, and L2 will remain at the same brightness; the power draw on the battery will double.

26. A simple parallel circuit with resistances of 2.5 Ω and 5.0 Ω, respectively, will have:
(A) equal amounts of current through each resistor
(B) twice as much current in the 2.5 Ω resistor
(C) twice as much current in the 5.0 Ω resistor
(D) 100 percent of the current in the 2.5 Ω resistor

27. Plugs are often differently shaped in different countries to prevent foreign electronics from drawing on an electric grid they weren't designed for. What would happen if a device designed for a 150-volt source were forced into a plug providing 240 volts?
(A) The device would function normally, as current and resistance are constant within a device.
(B) The device might be weaker due to a lower current.
(C) The device might burn out due to higher currents.
(D) The device would function normally but waste more energy.

28. Which one of the following circuit components will least affect the flow of an alternating current?

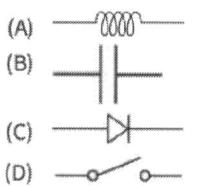

(A)
(B)
(C)
(D)

29. the fuse is rated at 0.4 A in the wiring diagram below. The lamps and resistor have the resistances given, and the fuse has no resistance. What will happen when the switch is closed?
(A) Both lamps will stay lit.
(B) Both lamps will light briefly and go out.
(C) L1 only will stay lit.
(D) L2 will only stay lit.

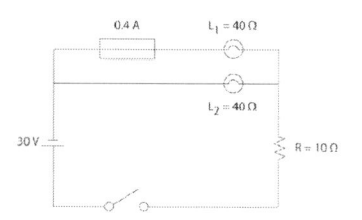

30. A technician wants to measure the total current in the circuit described in the wiring diagram. If she attaches ammeters just after the 10 Ω and 20 Ω resistors, the total current will be equal to:

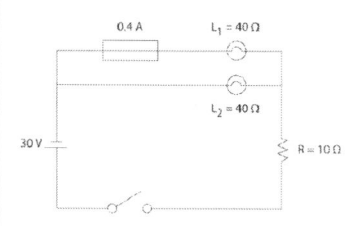

(A) the sum of the values on each ammeter
(B) the value on either ammeter since they'll show the same value
(C) the difference in the values on the two ammeters
(D) the average of the values on the two ammeters

Answers and Explanations
ELECTRONICS INFORMATION PRACTICE SET 1

1. **A**

To solve this problem, the first step is to convert 60 mV to V. You can quickly make this conversion by moving the decimal point three places to the left, which yields 0.060 V. Next, convert 15 kΩ to Ω by moving the decimal point three places to the right, which generates 15,000. Since the current is unknown, the formula needed to finish the calculation is $I = E \div R$. Dividing 0.060 by 15,000 gives 0.000004 amperes. This can be converted to 0.004 mA.

2. D
Like the last problem, this requires converting the given measurements first. Take the resistance and move the decimal. 60 kΩ becomes 60,000 Ω. Using our formula for calculating voltage, $E = I \times R$, we get 30 A × 60,000 = 1,800,000 V, or 1.8 MV.

3. B
The resistors in a series circuit add up, so one 40-ohm resistor in series with a 20-ohm resistor equals a 60-ohm resistor. To find current flow through a circuit, divide the total circuit resistance (60 ohms) by the voltage across the circuit (60 volts) to get 1 amp.

4. D
The valence shell is the outer shell of an atom, the fourth shell in a copper atom. A copper atom has 29 electrons, so it would fill the first three shells and have one electron left to begin a fourth shell.

5. B
In the symbol, the number 2 indicates the collector.

6. C
The ampere is the basic unit of current, so current can be expressed in Milliamperes.

7. B
A closed circuit has continuity and will allow current to flow in it.

8. C
The choices represent four different capacitor symbols: a fixed capacitor (A), variable capacitor (B), polarized capacitor (C), and trimmer capacitor (D).

9. B
When a P-type material is joined with an N-type material, a diode is Created.

10. D
Examine the diagram carefully. Notice that switches S1, S3, and S4 can all direct current flow to either of two alternative paths. Switch S2 will bridge the current from one of the top parallel paths to the other. The circuit pictured is currently open. The S1 and S3 two-switches connect to the same parallel paths, each connected to a different branch. If either of these switches were to flip, a complete circuit would be made, passing through one or the other of these paths. S2 is a simple on/off switch; if closed, it would close the circuit. However, S4, a two-switch, cannot affect the current break in the circuit.

11. A
This question has no numbers, which can make it tougher to grasp. It can be reasoned out if your understanding of parallel and series resistors is solid. Otherwise, your best bet might be to choose some simple numbers and do the math. The question has three identical resistors in a series-parallel circuit. Choose an easy number—three 1 Ω resistors would work just fine, and you can determine the correct answer by seeing if the effective resistance is greater than, less than, or equal to 1 Ω. The effective resistance of two 1 Ω resistors in parallel is ½ Ω. Add 1 Ω resistor in series, and the final effective resistance is 1 ½ Ω, an increase. Time permitting, the calculation can be repeated with a different resistance value for the three resistors to ensure that the resistance still increases.

12. C
Each answer choice would increase the rate of current flow, except one. Answer choices (A) and (D) would decrease the circuit's overall resistance, which, according to Ohm's law, would increase current. Answer choice (B) would also increase the current flow if resistance were not changed. However, an inductor in the circuit has little effect on a direct current. It will resist changes in current flow but not tend to either decrease or increase it.

13. B
The effective resistance of several resistors wired in parallel can be determined by taking the reciprocal of the sum of the reciprocals of each resistance value. Since each resistor has a resistance of 2 Ω, you get ½ + ½ + ½ = 3/2. Since the reciprocal of this sum is needed in parallel resistor calculations, it's best to stick with fractions at least until the end. The reciprocal of 3/2 is simply the fraction created by switching the numerator and denominator values, ⅔ Ω.

14. D
Copper isn't the best conductor. Silver has a lower resistance and is thus better at conducting electricity than copper. But though silver is a somewhat better conductor, it is far, far more expensive. Copper is a very good (but not the best) conductor and affordable.

122 | ASVAB Study Guide 2024-2025

Aluminum is inexpensive but is a worse conductor than copper. Copper is also safer than aluminum, which, due to some of its heating and expanding properties, is more likely to cause electrical fires in household wiring setups.

15. B
Rearranging the power formula, $I = P \div V = 10 \div 6 = 1.67$ A (rounded to the nearest hundredth).

ELECTRONICS INFORMATION PRACTICE SET 2

16. D
The relevant formula here is $P = I \times V$, and both the current and voltage are given, so $P = 30 \times 2.5 = 75$ W.

17. A
The relevant formula here, $V = I \times R$, must be rearranged to calculate resistance. $R = V \div I = 120 \div 15 = 8\ \Omega$.

18. C
These two resistors are parallel and can be treated as a unit. Since we're given the voltage drop across them both (the voltage drop either in total or across either resistor is the same when wired in parallel) and the total current through them both, we can calculate the total effective resistance of the two resistors. $R = V \div I = 18 \div 6 = 3\ \Omega$. The question asks for the resistance of either resistor, though, so there's one more step. The formula for the total effective resistance of these two resistors in parallel is $1/R_{tot} = 1/R_1 + 1/R_2$. However, since we know both resistors are the same and we know the total resistance, the formula can be written as $\frac{1}{3} = \frac{1}{R} + \frac{1}{R}$. R is the resistance of a single resistor and is exactly what we're trying to figure out. Multiply both sides of the equation by R to get $R/3 = 1+1$ or $R/3 = 2$. Then multiply both sides by 3 to get $R = 6\ \Omega$.

Alternatively, you might recall that in the special case of multiple identical resistors in parallel, you can use the shortcut of dividing the individual resistance by the total number of resistors. To have a total effective resistance of $3\ \Omega$, there must be two $6\ \Omega$ resistors ($6 \div 2 = 3$).

19. D
Household wiring and long-distance high-voltage power lines are connected via step-down transformers in the local substations. The other statements are not true. The current is AC in both power lines and household wiring; power lines do not always use aluminum wiring.

20. B
The first and second colors determine digits, and the last determines how many zeroes to add. Black is zero, orange (in either of the first two positions) is three, and orange (in the final position) is three zeroes. A zero at the beginning of the number has no effect, as 03000 is just written as 3,000. Answer choice (C) would have been correct if the pattern were orange, black, and orange.

21. A
Inductors resist changes in current and act as resistors in alternating current circuits. However, this property is also important in a direct current circuit, as seen here. An open circuit has no current, but the current will begin to flow when the switch is closed. The inductor, resisting the sudden voltage change, will first resist current flow. Since it is parallel to the resisting inductor, the lamp will light up immediately without difficulty. However, the inductor will soon "get used" to the current flow, and the resistance of its path will drop significantly. As more current flows through the easy inductor path, the circuit's total current (and energy use) will increase significantly as the resistance drops. (D) is precisely the opposite of what will happen. Something similar to (B) could happen only if the voltage source were limited in the amount of current it could provide. The small fraction of current passing through the lamp was insufficient to light it once the inductor dropped its resistance or if the total current draw caused a fuse or breaker to cut off the current. However, in neither of these cases would the lamp itself "burn out."

22. A
The hair dryer has the same resistance, no matter what circuit it's attached to. You can calculate the resistance from the specifications and then use that same resistance to determine the current under a lower voltage. $R = V \div I = 220 \div 10 = 22$ ohms. Plugging this 22-ohm hair dryer into an American socket gives you the current as $I = V \div R = 110 \div 22 = 5$ amps, which is answer choice (A). Another way to answer this question is by reasoning it out from the proportionality given by Ohm's law: $V = I \times R$. Given a constant resistance, voltage and current are constant, so if the voltage is halved from 220 to 110 volts, the current must also be halved from 10 to 5 Amps.

23. B
This question about power takes you back to the more general definition of *power*, as the energy rate flows over time. You do not need the electrical power formula, $P = I \times V$. Instead, divide the total energy in joules by the total time in seconds to determine the total power in watts. The information in the question is already given in the correct units, so calculate 800,000 ÷ 400 = 2,000 W.

24. D
You can dismiss (A) and (B) since, for loads in parallel, the voltage drop is the same across all paths; the brightness of the light would not change if the lamp were wired in parallel with either a capacitor or an inductor. Loads in series, on the other hand, affect the overall resistance of the circuit and, therefore, the total power draw on the battery. Recall that capacitors and inductors can let current flow freely or act as resistors, depending on the current type. In an AC circuit, a capacitor allows current to flow freely while an inductor resists current flow. An inductor placed in series with the lamp will, therefore, act as a resistor over which a voltage drop occurs, and the lamp will be correspondingly dimmer than if the circuit's voltage drop occurred entirely over the lamp itself.

25. D
Before the switch is closed, only one path through the circuit, through L2, is lit while the other lamp is dark. Closing the switch will provide a second possible path, so both lamps will now be lit. However, it's also necessary to determine what will happen to the brightness of L2 itself. The general rule is that lights in parallel will not change their brightness as more paths are added because the voltage drop across each path remains the same, and therefore, the amount of current passing through each lamp also stays the same. However, the fact that there are now two brightly lit lamps instead of one means that the battery will drain twice as fast because the total power draw of the circuit will be doubled when the switch closes.

26. B
$I = V ÷ R$, and since the voltage across parallel loads is always equal, any current difference will result from the different resistances. You can see that current and resistance are inversely proportional, meaning any change to one will have the opposite effect on the other. But this question may be a little easier to figure out by just making up a value for voltage, say 5 V. Given this value, the currents through each load will be $I = 5 ÷ 5 = 1$ A, and $I = 5 ÷ 2.5 = 2$ A. The 2.5 Ω resistor has twice as much current as the 5.0 Ω resistor.

27. C
Since the device's resistance is constant, the formula $V = I \times R$ shows that if the voltage increases, the current increases proportionally. Put another way, the greater the voltage, the greater the "pressure" pushing those electrons forward. This has the effect of producing a higher current, assuming the resistance is the same. Only answer choice (C) mentions the higher current that must result from this situation. And yes, you should only plug your device into a socket that fits so you don't burn it out or start a fire.

28. B
This is the general symbol for a capacitor, which acts as a resistor to direct current but does little to disrupt alternating current. It's the opposite of an inductor, pictured in answer choice (A), which acts as a resistor to alternating current. Choice (C) is the diode symbol, which allows current flow in only one direction. Since alternating current constantly changes direction, half of the current flow would be disrupted by the process known as rectification. Finally, choice (D) displays an open switch, which will not allow AC or DC power to Flow.

29. D
The total current in the circuit can be determined from the circuit's voltage source and total resistance. This is a tricky series-parallel circuit that has to be taken in two steps. The total resistance of the parallel loads L1 and L2 can be calculated without the full parallel-resistors formula by using the rule that equal resistors in parallel have an effective resistance equal to their resistance divided by the number of parallel loads. The effective resistance of L12 then is 40 ÷ 2 = 20 Ω. This is in series with R, so the total resistance of the whole circuit is 20 + 10 = 30 Ω. Now, calculate the current draw of the circuit as $I = V ÷ R = 30 ÷ 30 = 1$ A. Once again, the fact that L1 and L2 have equal resistance makes things easier. Since the two paths are equivalent, each will take half the current, meaning 0.5 A passes through either one. This is greater than the maximum current allowed by the fuse, so it will burn out, cutting the current flow through L1 and leaving only L2 lit.

30. A
Current will separate when reaching parallel paths. Since the two resistors are different, they'll draw different amounts of current, so the statement in (B) is false. However, regardless of whether they're the same or different, the sum of the two currents will give you the total current. However, this technician would have been better off placing her ammeter just after the battery or near the switch, as doing so would have allowed her to get the total current with a single measurement.

Chapter 11

Automotive and Shop Information (AS)

Engine Systems:

Internal combustion engines are the kind of engines found in cars. A mixture of air and fuel burns quickly during combustion. It sounds like internal combustion: gasoline is burned inside, and the heat produced directly drives the engine.

Regular fuels for internal combustion engines include alcohol, propane, diesel, gasoline, and natural gas. The most popular fuel types are, by far, gasoline and diesel. Air, fuel, and a heat source that can ignite the air-fuel mixture are the three components that must be present for combustion.

The engine won't run if any of the three components is lacking because that will stop combustion. Internal combustion engines transfer the chemical energy in the fuel and air into heat energy, which is subsequently transformed into mechanical energy.

Components:

There are standard components in all internal combustion engines. These components include:

- **The Engine Block** forms the framework for the engine cylinders and reciprocating assembly.
- **Piston:** a cylindrically shaped object with a solid crown (top) that moves up and down in the engine's cylinders. Hot gases produced from the combustion of the air-fuel mixture push on the piston to do the actual work.
- **Cylinder:** forms a guide for the piston to move in; allows the piston to move up and down as the engine completes its cycle.
- **Piston Rings:** seal the piston to the cylinder and prevent combustion gases from leaking past. Oil rings prevent oil from the engine crankcase from entering the combustion chamber.
- **Wrist Pin:** connects the piston to the connecting rod and forms a pivot point for the small end of the connecting rod to move on.
- **The Connecting Rod** connects the piston/wrist pin assembly to the engine's crankshaft. The large end of the connecting rod attaches to the crankshaft on the connecting rod journal.
- **Crankshaft:** converts the piston's linear (straight line) motion into rotary motion, which can then be used to power a vehicle or drive an accessory.
- **Cylinder Head:** located above the piston, houses the combustion chamber, the intake and exhaust valves, and the intake and exhaust ports.
- **Combustion Chamber:** situated in the cylinder head directly above the piston, it is where the actual combustion of the air-fuel mixture takes place.
- **Intake Valve:** draws the air-fuel mixture into the combustion chamber. When closed, it must seal the combustion chamber from the intake port.
- **Exhaust Valve:** allows waste gases to be removed from the combustion chamber. When closed, it must seal the combustion chamber from the exhaust port.
- **Camshaft:** responsible for the opening and closing of the engine's intake and exhaust valves. The camshaft turns at one-half the speed of the engine's crankshaft.

COOLING SYSTEM:

The two main cooling system types in contemporary automobile engines are as follows. The first kind is air-cooling, which involves blowing air over cooling fins on the engine's exterior to remove extra heat. Water cooling is the second kind.

A water-cooled engine uses a liquid coolant to absorb excess heat expelled through a radiator. Cylinder temperatures must promptly achieve operating temperature and hold steady over a wide range of ambient temperatures to guarantee improved emissions and efficiency.

As a result, liquid cooling is used in almost all newly developed automotive engines due to its higher flexibility in coolant circuit design, ability to transmit heat, and high heat capacity.

Coolant:

The coolant itself is the most essential part of the cooling system. Antifreeze and water are often blended 50/50 to make engine coolant. The coolant must be freeze-protected since frozen coolant can cause catastrophic engine damage, including the possibility of a broken block and/or cylinder head.

Ethylene glycol is the most widely used kind of antifreeze. Ethylene glycol and water mixed 50/50 will not freeze until it reaches -34 degrees Fahrenheit. The coolant's boiling point will be raised by the same 50/50 mixture, which is crucial in hot weather since it increases the coolant's capacity to transmit heat.

Components

The following are the main elements of a water-cooling system:

- **Water Pump**: Moves coolant through the cooling system to transfer and control heat.
- **Water Jacket**: Hollow sections in the engine block and cylinder head that allow coolant transfer. These are the areas where the coolant must absorb heat.
- **Thermostat**: Controls engine temperature by allowing coolant to flow into the radiator when the coolant temperature rises above a certain level.
- **Bypass Tube**: Allows coolant to flow back into the water pump from the cylinder head when the thermostat is closed.
- **Radiator Hoses**: Flexible hoses will enable hot coolant to flow between the engine and the vehicle's radiator.
- **Radiator**: Responsible for transferring heat from the coolant to the outside air.
- **Radiator Cap**: Responsible for maintaining pressure in the system and transferring coolant between the coolant reservoir and the radiator.
- **Coolant Recovery Bottle**: Forms a reservoir for coolant to flow in and out of the cooling system as the engine increases and decreases in temperature

A water pump driven by a belt or timing chain moves coolant throughout the engine. The engine powers the water pump since the engine crankshaft drives the belt. The water pump draws in and forces coolant into the engine block.

The coolant ascends into the cylinder head and travels back to the water pump's inlet through the bypass tube. The coolant will circulate in this way as long as the thermostat is closed (the engine is below operational temperature). Hot coolant enters the radiator itself after passing through the thermostat and entering the top radiator pipe when the thermostat opens.

Using coolants composed of a 50/50 mixture of ethylene glycol and water is one of the techniques used to enhance the boiling point of water, as was already indicated. An alternative strategy is to increase the cooling system's pressure. The radiator cap causes this, most of which are made to keep the cooling system's pressure between 9 and 16 pounds per square inch (psi).

As a result, the boiling point of the coolant is elevated by around 3°F for every 1 psi of pressure applied to the cooling system. The coolant's boiling point would then increase to between 212°F and 260°F with a 15-psi radiator cap.

A pressure valve and a vacuum valve are both built into radiator caps. The cooling system's coolant will expand as the engine warms up. When the radiator cap's rating is reached, the pressure valve in the radiator cap will lift, allowing some coolant to flow from the radiator into the coolant recovery bottle. This expansion will increase the cooling system's pressure.

This flow will continue if the cooling system pressure exceeds the radiator cap's rating. Until the engine reaches operating temperature, the coolant level in the recovery bottle will increase. The coolant will shrink when the engine is turned off and starts to cool, creating a low-pressure area inside the cooling system. If uncontrolled, this low-pressure zone would collapse the radiator hoses and negatively impact the cooling system when the engine starts.

Maintenance

For cooling systems to function at their best, care is necessary. Once more, replacing the coolant every two to three years is an intelligent maintenance practice because it is so important. A 50/50 mixture of antifreeze and water should maintain the coolant's strength. This will guarantee that the coolant has adequate corrosion resistance and freeze protection.

Additionally, it's crucial to make sure the system is leak-free and that the belt powering the water pump is in good shape. Inspecting the radiator and the hoses for apparent wear or damage is also recommended. A leak will usually be visible from the outside and drip from damaged cooling system parts or hose connections.

LUBRICATION SYSTEM:

Another system critical to engine operation is the **lubrication system**. Without lubrication, the engine's internal parts would quickly develop enough friction to stop (seize) the engine completely.

- The lubrication system is responsible for the following functions:
- **Lubricates:** It puts an oil film between moving parts to reduce friction and smooth engine operation.
- **Cools:** It puts motor oil in contact with hot engine parts (such as the underside of the piston) and transfers heat to the oil pan or the engine oil cooler if applicable.
- **Seals:** Motor oil acts as a sealer between the piston, the piston rings, and the engine cylinder walls. This helps seal combustion gases in the combustion chamber and makes the engine run more efficiently.
- **Cleans:** Additives in the motor oil cause contaminants to be suspended so that the engine's oil filter can filter them out.
- **Quiets:** Motor oil damps engine noise and makes the engine run more quietly.

Engine Oil:

A significant part of the lubricating system is engine oil. The lubrication for the engine's moving parts is engine oil. The longevity of an engine depends heavily on the engine oil, which must be adequately developed to deliver top performance under the worst engine operating circumstances. The two primary engine oil parts are the base oil and an additive package.

Viscosity is one of the most crucial characteristics of engine oil. Viscosity, defined as the resistance to flow, is quantified as a value inversely proportional to the oil's thickness. The specifications for engine oil viscosity were created by SAE or the Society of Automotive Engineers.

For example, an oil with an SAE 5 viscosity rating would have a low viscosity (low resistance to flow). In contrast, an oil with an SAE 50 viscosity rating would have high viscosity. Therefore, the. The API quality rating is a further significant engine oil evaluation system.

The American Petroleum Institute, or API, is in charge of establishing criteria for the caliber of motor oil. This rating would begin with the letter "S" for a gasoline engine, and the following letter would indicate the precise quality standard that the engine oil complies with. Greater engine oil demands resulted in ratings at the SN, SM, SL, and SJ levels.

The first gasoline engine oils produced had a quality grade of SA. If modern engines were run on SA motor oil, they wouldn't survive very long. Diesel engine use would be appropriate for engine oil with a "C" prefix for its quality rating (for example, a CJ-4, CI-4 plus, CI-4, and CH-4 grade). Using either gasoline or diesel engines with both an "S" and a "C" rating would be appropriate for engine oil. Engine oil with an SJ/CD rating would illustrate this.

Components:

The primary components of the lubrication system are as follows:

- **Oil Pan:** Forms the reservoir for the engine oil at the bottom of the engine.
- **Oil Pickup Tube and Screen:** Immersed in engine oil, this filters out large solids and directs oil into the pump.
- **Oil Pump:** Responsible for pumping the oil through the engine oil galleries. The engine's crankshaft typically drives it.
- **Pressure Relief Valve:** Prevents excessive pressure from building in the lubrication system.
- **Oil Filter:** Filters oil from the oil pump before sending it to the engine's various parts.
- **Oil Galleries:** Passages or "drillings" in the engine assembly that transport oil to critical components.

OPERATION:

Wet and dry sump lubrication are the primary designs for lubrication systems. Due to its complexity and limited application on racing vehicles, the dry sump system is infrequently used. Wet sump lubrication systems are used in all mass-produced commercial and automotive engines. The oil pan at the engine's base serves as the engine's oil reservoir in a wet sump system design. Due to gravity, all engine oil drains into the oil pan.

There is a chance for it to cool. The pump draws oil from the pan into its inlet through the pickup tube and screen. The engine oil is sent into the filter after passing through the oil pump. The major oil galleries get the clean oil after the oil filter cleans up the particles in the oil.

It takes much more energy to pump the oil through the system when the engine is cold because the oil has a higher flow resistance. Additionally, this raises the oil pressure in the engine. The pressure relief valve will open if the oil pressure exceeds the setting, allowing some of the oil to drain back into the oil pan before it is poured into the oil filter.

By doing this, engine oil pressure is kept from increasing to a level that could harm parts like the oil filter, bearings, and engine seals. The engine's reciprocating assembly is the most crucial component of lubricating needs. Oil is necessary for the crankshaft, connecting rods, and pistons to reduce friction and dissipate heat. To provide oil under pressure to the connecting rod bearings, galleries are immediately bored through the crankshaft and used to transfer oil to the largest oil galleries in the engine.

The oil that seeps from the connecting rod bearings is discharged into the cylinder walls to lubricate and cool the pistons. Other oil galleries divert oil from the primary oil gallery to the engine's valve train. The rocker arms, valve stems, push rods, lifters, and camshaft are lubricated here.

The camshaft drive, timing chain, and related gear drives for the oil pump, ignition distributor, etc., could also be lubricated by this oil. After passing through the engine, the oil finally returns to the oil pan. Since the oil pan is near the bottom of the engine, it lies deep in the engine compartment, usually where cooler air is found.

This cool outside air assists in cooling the engine oil by absorbing heat from the oil pan. A liquid-to-air heat exchanger (like the radiator in the cooling system) can further cool the engine oil during heavy-duty or hot-weather operation. The oil is prepared to be picked up by the oil pump and pumped back into the lubrication system after it has cooled.

Combustion Systems:
FUEL SYSTEM:

Maintaining the ideal air-fuel ratio is the responsibility of the fuel system for effective engine running. However, driving comfort, emission management, fuel efficiency, and engine life may all suffer from improper air-fuel mixture adjustments.

Mechanical techniques were the most typical method for determining the air-fuel mixture until roughly 35 years ago. A carburetor was used to achieve this, which was very dependable but lacked the precision needed in modern fuel systems.

The carburetor has mostly been replaced by electronic fuel injection as a result of the requirements of emission control rules. This system has an inbuilt computer that controls it and a feedback feature that allows it to adapt quickly to shifting engine conditions.

Components:

The following are the major components of an electronic fuel injection system:

- **Electric Fuel Pump:** Located in the vehicle's fuel tank, it supplies fuel under pressure to the fuel injectors.
- **Fuel Filter:** Filters contaminants from the fuel before it reaches the fuel rail.
- **Fuel Rail:** A manifold that supplies fuel under pressure to the inlets of all the engine's fuel injectors.
- **Fuel Pressure Regulator:** Regulates pressure in the fuel rail according to the intake manifold vacuum. Excess fuel is bled to the fuel return line and sent back to the fuel tank.
- **Fuel Injector:** Sprays fuel into the intake air stream as it receives electrical signals from the powertrain control module (PCM). The location of the injector will determine the specific type of fuel injection system being used on the engine.
- **Powertrain Control Module (PCM):** Another name for the vehicle's central computer. Responsible for control of all functions associated with the engine and transmission.
- **Intake Manifold:** Distributes air to the intake ports on the cylinder heads.
- **Intake Air Filter:** Removes airborne contaminants that could damage internal engine parts. All air entering the engine passes through the air filter.
- **Throttle Body/Throttle Plate:** Connected to the throttle pedal; controls engine speed and output torque.

Fuel Injection System Designs:

When fuel injection systems were originally widely used, one or two injectors installed in a throttle body that replaced the carburetor were intended to be used. A throttle body injection system, or TBI, is what this is. Although the system proved dependable, it could not provide a high enough level of fuel control to meet the standards for pollution control.

Most engines are now constructed with direct injection, where the fuel is delivered directly into the combustion chamber, or multiport fuel injection, which has an injector for each engine cylinder. The injectors for multiport fuel injection are housed in the intake manifold, with the intake valves receiving the spray from the injectors. Before fuel is injected into the cylinder head, air will travel from the intake manifold to the head of the engine.

As a result, the air-fuel mixture is greatly improved, and fuel droplets are kept from landing on the intake manifold runners. When using direct injection, a high-pressure fuel injector is positioned to spray highly pressurized fuel directly into the combustion chamber.

As a result, only air enters the combustion chamber through the intake valves. Fuel can then be sprayed straight into the combustion chamber at an ideal time during the compression stroke. In addition, the gasoline serves as a coolant for the hot, compressed air inside the combustion chamber, enabling the engine to be designed with even higher degrees of compression, which boosts power and efficiency.

Maintenance:

The fuel injection system requires very little maintenance. The filters (both air and fuel filters) must be changed periodically based on the manufacturer's recommendations. Manufacturers may also require periodic cleaning of the injection system. Aside from that, fuel injection systems are built to give reliable service with minimum maintenance. IGNITION SYSTEM

One of the most critical vehicle systems is the **ignition system**. Without it, combustion cannot occur, so the engine will not run. The ignition system must generate high-voltage sparks at the correct time to make the engine run smoothly and efficiently.

Components:

The ignition system can be divided into two distinct subsystems: the primary ignition system and the secondary ignition system. The primary is the low-voltage part of the system, whereas the secondary is high voltage.

The following are the major components found in the **primary ignition system** of a basic electronic ignition system:

- **Battery:** Supplies power to the ignition system for starting the engine.

- **Ignition Switch**: Turns the engine on and off by switching power to the ignition system.
- **Primary Coil Winding**: The low-voltage winding in the ignition coil. This is made up of several hundred turns of relatively heavy wire.
- **Ignition Module**: A transistorized switch that turns the primary current on and off.
- **Reluctor and Pickup Coil**: Responsible for generating a signal that operates the ignition module. The reflector is mounted on the distributor shaft and generates a signal in the stationary pickup coil as it rotates with the distributor.
- **Distributor**: Driven by the engine's camshaft, the distributor is responsible for timing the spark and distributing it to the correct cylinder. Like the camshaft, the distributor turns at one-half the engine's speed.

The following are the major components found in a basic electronic ignition system's secondary ignition system.

- **Secondary Coil Winding**: The high-voltage winding in the ignition coil. This comprises several thousand turns of fine wire wound around the primary coil winding.
- **Coil Wire**: Transmits high voltage from the secondary coil winding to the distributor cap.
- **Distributor Cap and Rotor**: Directs high voltage from the coil wire to each cylinder in the firing order. This switching mechanism allows one ignition coil to serve all the engine cylinders.
- **Spark Plug Wires**: Transmit high voltage from the distributor cap to each spark plug.
- **Spark Plugs**: Generate the spark to initiate combustion. They are threaded into the cylinder head and protrude into the combustion chamber.

Primary Ignition Operation:

Current is transferred from the battery via the ignition switch and onto the primary coil winding when the driver turns the ignition switch to the "run" position. That identical current leaves the coil winding and travels via the ignition module and the vehicle ground circuit before returning to the battery. As a whole, the low-voltage (primary) system of the ignition system is in charge of managing the high-voltage (secondary) system.

Everything in the secondary system results from what occurs in the primary system. For example, the electromagnetic induction phenomenon makes the ignition system function. Voltage is induced in a stationary wire when a magnetic field passes over it.

The creation of a magnetic field surrounding a wire due to an electric current flowing through it is another aspect of this theory. This indicates that an electric current can create a magnetic field, and a magnetic field can create an electric current.

Secondary Ignition Operation:

A high voltage is induced in the secondary winding by the primary coil winding's magnetic field collapsing. As a result, modern systems can have considerably higher secondary system voltages, typically ranging from 20,000 to 40,000 volts.

Because there are many more wire turns in the secondary winding than in the primary, a significant boost in voltage is achieved in the ignition coil. A step-up effect is produced in the coil due to the magnetic field in the main winding cutting across a significantly larger number of wire turns in the secondary.

The proper cylinder in the firing order must receive the high-voltage current from the secondary winding. The middle tower of the distributor cap receives this current through the coil wire. From there, it travels to the rotor, which rotates with the distributor and is positioned atop the shaft. When the engine turns, the rotor will route the current to the proper cylinder, sending it through the spark plug wire to the spark plug. The spark plug's electric arc is what starts combustion.

Current Ignition System Design:

Engineers carefully consider how to remove more moving parts from the ignition system since engine performance strongly depends on precise ignition timing. Due to this, the distributor was eliminated in a design known as the distributor-less ignition system, or DIS.

A V-8 engine would need four independent ignition coils instead of one current DIS system, which employs one ignition coil to power two spark plugs. The more coils there are, the more time each coil has to build up its electric charge and produce a stronger spark when it's time to discharge.

To function effectively, the DIS system still uses spark plug wires, which deteriorate over time and eventually need to be replaced. Recent ignition system designs completely do away with spark plug wires. Coil-on-plug ignitions are the name given to these systems.

The computer system in the car regulates the discharge of the individual ignition coils, which are located right on the spark plugs. While the addition of computer control has increased the ignition system's complexity, removing moving and high-maintenance components has resulted in highly accurate spark timing control, which is essential for optimum engine efficiency and emissions and increased ignition system reliability.

EXHAUST SYSTEMS:

The exhaust system is responsible for removing waste gases from the engine. It must do this in a way that allows these gases to flow freely, muffles the sound of the exhaust, and keeps the gases and heat away from the vehicle cabin and its occupants.

Components:

Typical components found in an exhaust system include the following:

- **Exhaust Manifolds**: Attached directly to the exhaust ports on the cylinder head. Most of the exhaust heat and noise is focused on the exhaust manifolds. These are often made from cast iron for durability under high heat conditions.
- **Catalytic Converter**: This converter is responsible for converting the toxic components of engine exhaust into relatively harmless compounds such as carbon dioxide and water.
- **Muffler**: Incorporates an expansion chamber and sound-absorbing material to diminish loud exhaust noises.
- **Tailpipe**: The exit point for exhaust gases as they enter the open atmosphere. The tailpipe usually exits at the rear of the vehicle.

Operation:

The exhaust manifold gathers gases as they come out of the engine's exhaust ports. There would be two exhaust manifolds on a V-type engine, one for each cylinder head. The major parts of the exhaust system are connected by steel exhaust pipes fed by these manifolds.

The catalytic converter is then filled with the exhaust gases. The catalytic converter generates a lot of heat as it transforms the hazardous gases in the exhaust into less toxic ones. Some cars employ two or even three catalytic converters to comply with emission control laws.

Gases from the exhaust are sent into the muffler after leaving the catalytic converter—the expansion chambers in the muffler help to dampen the loud noises produced by the engine's combustion. The muffler is often found toward the back of the vehicle, between the catalytic converter and the exhaust pipe. Engine exhaust is still very hazardous even after it has gone through the catalytic converter.

The most hazardous of the harmful gases released by an engine's exhaust is carbon monoxide, a deadly gas with no smell or color. Therefore, these gases must be transported without touching the driver or other passengers in the car. This often includes venting these gases out the tailpipe and into the open air at the back of the car.

The exhaust system can be tailored to the engine's design to boost efficiency and power. Header pipes are another name for "tuned" exhaust manifolds. The goal is to improve engine breathing by allowing exhaust gases to flow more freely. However, since header pipes are far more brittle than cast-iron exhaust manifolds, increased performance is typically at the expense of durability.

Electrical and Control Systems:
ELECTRICAL SYSTEM:

With each new model year, the importance of the vehicle's electrical system increases. Many mechanically driven systems are currently being rebuilt to run electrically.

Subsystems:

The following are some of the electrical system's primary subsystems: Batteries chemically store electrical energy. Provides direct current (DC) for accessory functioning and engine starting. The starting system is in charge of turning the engine over to start it. The two main parts of the starting mechanism are the battery and the starter motor. The charging system supplies the electrical current needed to refuel the battery and power the car.

The alternator is a vital element of the charging system. The lighting system includes headlights, turn signals, brake lights, and taillights. The accessories category includes the blower motors, audio, rear-window defogger, windshield wipers, and all other electrically operated items.

Battery:

The basis of the entire electrical system is the battery. When the load exceeds the alternator's output, it supplies current to the electrical system and serves as an electrical "shock absorber," reducing voltage spikes when it draws too much current. It also provides electrical current for starting the engine. Lead plates submerged in sulfuric acid and water-electrolyte make up a vehicle battery.

Because of this, this kind of battery is referred to as a lead-acid battery. When the battery discharges, the electrolyte's sulfuric acid is converted to water, and the lead plates turn into lead sulfate. The lead plates and the electrolyte's chemical makeup are restored during battery charging. Because a vehicle battery contains a significant quantity of electrical energy and very corrosive sulfuric acid, care must be given when handling it.

Starting System:

An electrically powered starting motor is used to start the engine. The starting solenoid receives an electrical signal when the ignition switch is turned to the "start" position, engaging the starter drive gear onto the engine's ring gear (found on the flywheel). The starter motor rotates the engine quickly enough to start it once the solenoid connects the battery to it when the drive is engaged.

Charging System:

When the engine is operating, the charging system supplies electricity to operate the car's electrical system and recharge the battery. The alternator is the central element of the charging system. The engine's crankshaft drives a belt-driven alternator, transforming mechanical energy into electrical energy.

Alternating current (AC) is generated by the alternator and is then rectified by an internal bridge of diodes known as the rectifier. Direct current (DC), which can be utilized to power the car's electrical system, is created by the rectifier bridge from alternating current (AC). A voltage regulator manages the output of the alternator. During engine running, the typical system voltage is about 14.5 volts.

The electrical system becomes more loaded, and the voltage decreases when the vehicle's headlights, heater, and other accessories are turned on. As a result of the voltage regulator detecting this drop in system voltage, the alternator's output is increased to compensate for it. The electrical system voltage will stay close to 14.5 volts as long as the alternator's output can support the load.

Lighting System:

A vehicle's lighting system includes numerous lights. When appropriate, interior lights let the driver see the instrument panel and other inside spaces, while headlights and taillights assist other drivers in seeing the car's rear. The driver operates switches that turn on and off the electrical current to the lighting circuits to control the numerous lights.

Fuses or circuit breakers shield the lighting circuits in the vehicle. If the circuit draws more current than intended, the fuse will "blow" and stop the current flow. This shields the circuit wires from damage and may even prevent an electrical fire from starting.

COMPUTER SYSTEM:

Like the human nervous system, computer control systems operate similarly. The computer receives sensor data, analyses it, and then sends signals to actuators to control vehicle functions. In the human body, the brain processes information from the eyes and hearing and then sends signals to the various muscles to regulate movement.

Chassis Systems:
DRIVETRAIN SYSTEM:

Although the engine can provide enough power to move the car, smooth and rapid acceleration requires efficient energy processing and transmission. The vehicle's drivetrain is in charge of transferring power from the engine to the wheels. Therefore, the transmission is a crucial component of any drivetrain. The mechanism that adjusts the engine speed to the required vehicle speed is known as the transmission.

The automatic and manual transmissions are the two primary varieties. Because the driver is in charge of shifting the gears up and down, manual transmissions are frequently chosen by drivers who prefer to have greater influence over how the car operates.

An automatic transmission car is significantly simpler because there is no driver-operated clutch, and all gear shifting is done automatically. While front-wheel drive has gained popularity, especially in small to medium-sized cars, rear-wheel drive used to be the most frequent configuration.

SUSPENSION AND STEERING SYSTEM:

The newest suspension and steering systems have incorporated computer control to improve vehicle performance. However, while computer control's sophistication has increased these systems' overall complexity, the basic principles remain the same.

Components

The major components of the long-short-arm suspension system include the following:

- **Springs:** Hold the vehicle's chassis up and allow the wheels to move up and down in relation to it.
- **Shock Absorbers:** Absorb the energy released by the up-and-down movement of the vehicle wheels.
- **Control Arms (A-arms):** The long-short arm suspension system uses an upper and lower control arm to maintain the vertical orientation of the steering knuckle as the wheel moves up and down.
- **Steering Knuckle:** Connects to the upper and lower control arms through ball joints—the wheel hub mounts on the spindle, which is part of the steering knuckle.
- **Ball Joints:** Ball-and-socket assemblies that allow the steering knuckle to simultaneously turn and move up and down.
- **Steering Linkage:** Connects the steering wheel to the steering knuckle.
- **Wheel Hub:** Forms the mounting point for the vehicle's tire assembly.
- **Tire:** Makes contact with the road and provides a "footprint" to aid vehicle stability and handling.

BRAKE SYSTEMS:

The most important of all the systems on a car may be the brake system. It is one thing not to be able to move forward; it is another thing again to be unable to stop.

Components

The major components found in any brake system include the following:

- **Brake Pedal:** The mechanical connection between the driver's foot and the master cylinder.
- **Master Cylinder:** Located in the engine compartment just in front of the driver, the master cylinder generates the fluid pressure to operate the brake assemblies at the wheels.
- **Fluid Reservoir:** Provides fluid to the brake circuits. The fluid reservoir is located on top of the master cylinder.
- **Brake Lines:** Transmit fluid pressure from the master cylinder to the brake assemblies. These can be steel lines that run along the vehicle chassis or flex hoses that connect the steel lines to the brake assemblies.

Auto Information Practice Set 1

1. How many crankshaft revolutions does it take to complete one cycle of events in a four-stroke cycle engine?
(A) one-half
(B) one
(C) two
(D) four

2. This image depicts what stroke in the four-stroke cycle?

(A) intake stroke
(B) compression stroke
(C) power stroke
(D) exhaust stroke

3. A 12:1 air-fuel mixture would be a mixture.
(A) rich
(B) lean
(C) stoichiometric
(D) none of the above

4. Diesel engines utilize a:
(A) high compression ratio
(B) spark ignition system
(C) low compression ratio
(D) carburetor

5. The antifreeze in the engine's coolant will:
(A) lower the boiling point of the coolant
(B) raise the boiling point of the coolant
(C) raise the freezing point of the coolant
(D) lower the corrosion resistance of the coolant

6. Identify this engine component.

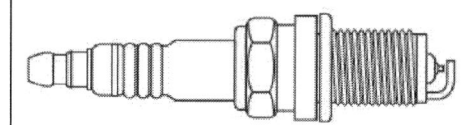

(A) spark plug
(B) fuel injector
(C) wrist pin
(D) valve spring

7. Electronic fuel injectors are controlled by the:
(A) powertrain control module
(B) fuel pressure regulator
(C) electric fuel pump
(D) ignition switch

8. Where does combustion take place in the engine?

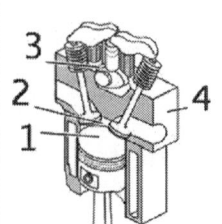

(A) 1
(B) 2
(C) 3
(D) 4

9. The PCM receives signals from the vehicle:
(A) sensors
(B) diagnostic data link
(C) actuators
(D) scan tool

10. Oil galleries in the crankshaft provide pressurized oil to:
(A) the pistons
(B) the push rods
(C) the valves
(D) the connecting rod bearings

Auto Information Practice Set 2:

11. How do shock absorbers dissipate suspension system energy?
(A) by converting motion energy into electric power
(B) by converting motion energy into potential energy
(C) by converting motion energy into heat energy
(D) by neutralizing chemical energy

12. The water-cooled engine component that rejects heat is the:
(A) water pump
(B) thermostat
(C) radiator
(D) antifreeze

13. Multiport fuel injectors are located in:
(A) the throttle body
(B) the combustion chamber
(C) the intake manifold
(D) the flame front

14. Which of the following is developed in the catalytic converter?
(A) toxic gas
(B) high heat
(C) hydrocarbon gas
(D) carbon trioxide

15. Coil-on-plug ignition systems eliminate the need for:
(A) secondary coil winding
(B) spark plug wires
(C) spark plugs
(D) primary coil winding

16. The main component of an engine's charging system is the _____.
(A) power charger
(B) battery
(C) rectifier
(D) alternator

17. Automatic transmissions do not require a clutch because a _____ transmits power from the engine to the transmission.
(A) half-shaft
(B) torque converter
(C) drive axle
(D) constant-velocity joint

18. Which of the rack and pinion gears are found in which of the following applications?
(A) steering system
(B) manual transmission
(C) lubrication system
(D) automatic transmission

19. The most important property that enables a hydraulic brake system to transmit pressure from the master cylinder to all individual brakes is _____.
(A) viscosity
(B) incompressibility
(C) acidity
(D) adaptability

20. Which of the following is the least expensive and most effective maintenance procedure to maintain maximum engine life?
(A) lubricating all fittings
(B) regularly changing automatic transmission fluid
(C) regular ng the thermostat

Answers and Explanations
AUTO INFORMATION PRACTICE SET 1:

1. **C**
It takes two revolutions of the crankshaft to complete each four-stroke Cycle.

2. **C**
This image depicts the power stroke, the third stroke in the four-stroke cycle. The power stroke generates the engine's power.

3. **A**
A 12:1 air-to-fuel ratio would be a rich mixture, fast-burning and possibly producing black smoke.

4. **A**

Diesel engines operate with a high compression ratio. The compression ratio for a diesel engine can range anywhere from 16:1 to 22:1.

5. **B**

Antifreeze raises your coolant's boiling point, making it more efficient at transferring heat.

6. **A**

This is a spark plug. Spark plugs are threaded into the cylinder head, protruding into the combustion chamber and generating the spark to initiate combustion.

7. **A**

Of the items given, only the powertrain control module is responsible for firing and regulating electronic fuel injectors.

8. **B**

The combustion chamber, indicated in the diagram by the number 2, is where the actual combustion of the air-fuel mixture takes place. It is located in the cylinder head directly above the piston.

9. **A**

As the car's " brain, " it is the job of the PCM to gather signals from the car's sensors and generate outputs to control the vehicle's functions. The diagnostic data link, choice (B), connects a scan tool to the PCM; the scan tool (D) is what technicians use to read the PCM.

10. **D**

The oil galleries are drilled directly through the crankshaft to provide pressurized oil to the connecting rod bearings. The pistons, push rods, and valves do not receive pressurized oil from the crankshaft oil galleries, so (A), (B), and (C) are incorrect.

AUTO INFORMATION PRACTICE SET 2

11. **C**

The shock absorber provides damping to the suspension system. The damping effect converts motion energy into heat, dissipating through the shock absorber body into the atmosphere. The shock absorber does not generate electrical or potential energy or neutralize chemical energy. Therefore, choices (A), (B), and (D) are incorrect.

12. **C**

The radiator is the engine component designed to reject heat in a water-cooled engine. The other answer choices are incorrect because antifreeze is a subcomponent of the coolant; the water pump circulates the coolant, and the thermostat regulates the coolant temperature.

13. **C**

Throttle body fuel injectors are located in the throttle body, so (A) is incorrect. Direct injection fuel injectors are located in the combustion chamber, so (B) is incorrect. The flame front is the burning portion of the fuel-air mixture in the combustion chamber, so (D) is incorrect. Multiport fuel injectors are in the intake manifold, so (C) is correct.

14. **B**

The function of the catalytic converter is to further reduce unburned hydrocarbon fuel in the exhaust gas by decomposing the unburned hydrocarbon fuel in the exhaust gas into carbon dioxide, water vapor, and heat. As a result, high heat is developed in the catalytic converter.

Answer choices (A), (C), and (D) are incorrect because they represent components not developed in the catalytic converter.

15. **B**

Coil-on-plug ignition systems eliminate the need for spark plug wires because the ignition coil is mounted directly over the spark plugs. However, answer choices (A), (C), and (D) are still necessary components in a coil-on-plug ignition system and therefore are not correct.

16. **D**

The alternator converts mechanical energy into electrical energy needed to operate the vehicle's systems and to charge the battery as required. Power charger (A) is a generic term, not a specific component. The battery (B) stores electrical energy, and the rectifier (C) converts alternating current from the alternator to direct current.

17. **B**

A torque converter converts the rotational power of the engine to fluid power. A half-shaft (A) connects the transaxle to the drive wheels in a front-wheel drive vehicle. The drive axle (C) sends power from the differential to the drive wheels. Constant-velocity joints (D) are between the transaxle and front drive wheels.

18. **A**

Rack and pinion gears convert rotational motion from the steering wheel to the linear motion needed to change the wheels' direction. This type of gear arrangement is not used in any other systems listed.

19. **B**

If brake fluid were compressible, the fluid displacement created by the master cylinder would be diminished at the individual brake piston assemblies, and braking power would be severely diminished. Viscosity (A) refers to how easily a fluid flows. While this could affect brake performance, it is not the most important property. Acidity (C) is irrelevant to transmitting pressure and fluid movement. Adaptability (D) is not a relevant property.

20. **C**

Proper lubrication of internal moving parts is essential to long engine life. Excessive wear on internal engine parts can occur when the engine oil becomes dirty or experiences degradation of its lubricating properties. Internal combustion engines are lubricated by circulating oil internally, not by lubricating fittings (A). Changing the automatic transmission fluid (B) does not need to be done frequently and is more critical to the transmission life than the engine.

Thermostats (D) generally don't require maintenance; they are usually replaced if they fail to function correctly.

Chapter 12

Mechanical Comprehension

A Review of the Physics of Mechanical Devices:

Mechanics is the area of physics that encompasses the laws of motion, energy, and forces. Velocity, momentum, and Newton's second law relating force, mass, and acceleration are all mechanical concepts. All mechanical devices, including the simple machines discussed in this chapter, are based on the application of force to achieve a movement or change in the position of a mass.

MASS AND FORCE:

Everything has mass. The total amount of matter in an object is quantified by its mass. Larger things often have more mass than smaller ones, though this isn't always the case because some materials have more mass per volume than others (are denser). Mass is not a vector quantity because it has magnitude but no direction, in contrast to the force of weight, which will be covered in more detail later in the chapter.

The most crucial fact regarding mass for mechanics is that it relates to the amount of force necessary to obtain a specific acceleration. In other words, the harder it is to change something's motion, the more mass it has. The resistance of an item to changes in motion is frequently referred to as inertia, but inertia is simply a characteristic of mass. A push or a pull is a force.

The world is filled with forces. A baseball being hit into the stands, a tractor driving a plow, or a person pushing past others in a crowded store are a few examples of some of the more visible ones. We frequently take other forces for granted. The weight (force) that Earth's gravity exerts causes things and people to remain on the ground instead of floating in midair. Since all forces are vector variables, they exert their force in the same direction. The overall force acting on an object is called its net force.

Objects would not move (or, if they were already moving, they would continue moving without stopping) if no force were applied. Large forces are needed to change the velocity of heavy objects (like a freight train). All mechanical technology aims to apply forces to the movement of various mass quantities as efficiently as possible.

NEWTON'S LAWS OF MOTION:

All mechanical devices operate under the assumption that Newton's laws are true. Newton's first law describes the unchanging condition of motion of an item when it encounters no net force. Only when the linked ropes are strong enough to oppose the force of gravity drawing the bulk downward completely is a large container suspended by a pulley able to resist falling to the ground.

The formula $F = ma$, where m is expressed in kilograms (kg), and is expressed in meters per second squared (m/s2), and N is expressed in newtons (N), can be used to express Newton's second law. This rule expresses the linear relationships between the force needed to move an object relative to its mass or the desired acceleration.

In the past, one of the most significant functions of machines was to increase the force that a person or work animal exerted to move extremely large things that would otherwise be immobile. Ancient buildings like Stonehenge and the Egyptian pyramids are proof of a long history of human use of mechanical labor assistance.

According to Newton's third law of motion, every action has an equal and opposite response. All uses of force, including those made possible by straightforward machinery, are subject to this law. Only because a daisy chain of action-reaction force pairs from one piece of rope to the next carries the force of tension through a rope, making it feasible to pull on a rope and pulley to lift a weight.

Common Types of Forces:
GRAVITY AND WEIGHT:

We have already established that weight and mass are distinct. Weight is the force that gravity applies to an object, whereas mass is a feature of matter that exists intrinsically. One particular instance of gravity at work is the downward force that the Earth applies to objects outside.

The direction of weight for all objects on Earth is the same: straight down toward the center of the Earth. Therefore, weight may be measured simply using a spring scale, like in many people's restrooms. Generic Science provided the generic formula for Newton's Law of Universal Gravitation. It can explain why a human will weigh less on a planet of different sizes and masses (for instance, astronauts walking on the Moon only weighed as much as they did on Earth).

However, it is reasonable to presume that the place is here on Earth for routine mechanical labor until you are told differently. Without accounting for air resistance, the acceleration caused by gravity (g) for objects falling free at the Earth's surface is estimated to be roughly 9.8 m/s2. (Note that "near the Earth's surface" refers to anywhere on Earth, even the summit of Mount Everest or when parachuting from a plane that is hundreds of meters in the air. A stair ladder's top doesn't greatly affect the value of g.)

When applying Newton's second law, F = ma, to the particular situation of weight, the more generic symbol for force, F, is swapped out for the more precise force of weight, W, and the constant value of acceleration due to gravity, g, is used in place of a. As a result, the formula W = mg is modified, with W now equal to weight in newtons (N), m equal to the mass in kilograms (kg), and g equal to the acceleration of gravity (either m/s2 or N/kg), depending on the case. Therefore, multiplying an object's mass in kg by 9.8 will yield its weight in N.

FRICTION:

Several responsive forces act to resist movement. When you move an object across a surface, the rubbing of the two surfaces results in **kinetic friction**, which always acts in a direction opposite to the object's motion. Because it always opposes motion, kinetic friction will eventually slow any moving object to a stop. Ice is polished to minimize kinetic friction, but unless it is constantly being pushed, a hockey puck will eventually slide to a halt in even the most slippery of rinks.

When an object is not moving, a force attempting to slide it across a surface will encounter a different friction type. Attempting to push a stationary box across a concrete floor will require a person to overcome the **static friction** that opposes any attempted movement.

Both kinds of friction are based on two factors: the) **coefficient of friction**, μ (the Greek letter mu), which represents how much two materials resist sliding against each other, and the) **normal force**, *F*N, which represents the equal and opposite force a surface exerts when an object presses against it. The formulas for kinetic and static friction, respectively, are

$$F_f = \mu_k F_N$$
$$F_f = \mu_s F_N$$

Ff stands for the force of friction, and μk and μs for kinetic and static friction coefficients. The coefficients of friction are constant for any two surfaces in contact. The coefficients of friction are dimensionless quantities with no units of their own. Therefore, whatever force unit is used to measure *F*N (N or lbs, for example) will also be used to measure *F*f. Slippery surfaces like ice or wet asphalt have low coefficients of both kinetic and static friction with most objects.

Friction is directly proportional to these coefficients. That's why drivers of cars need to understand that weather conditions can make braking to a stop (which relies on the friction of the wheels with the road surface) less effective.

For any given surface, the coefficient of static friction is higher than that of kinetic friction. This means that more force is required to move a stationary object initially than to keep it moving afterward. Once the large static friction force has been overcome and the object is in motion, the smaller coefficient of kinetic friction takes over, and the resistant force of friction decreases. This is why moving heavy objects usually involves one extra big push at the beginning but a somewhat smaller force for the rest of the trip.

One more important difference between static and kinetic friction is that static friction only arises in response to an attempt to move an object along a surface. Like the) normal force that holds up objects with weight, it responds to an applied force by matching and counteracting it. The formula determines the maximum of a variable value for static friction, beyond which it breaks, and an object will begin to move.

Both kinds of friction forces are directly proportional to the normal force of the surface. As an experiment, try laying your palm flat and sliding your hand along a nearby surface like a table or desk. Feel the resistance of the movement due to kinetic friction. With your palm still flat, try pushing harder against the same surface while still trying to slide your hand along it.

Is it more difficult? If you push very hard against the surface, you may find you cannot even overcome the force of static friction and slide your hand without decreasing the normal force first. When a person leans against a wall, she allows a portion of the weight to push against the wall, which pushes back with an equal and opposite normal force. This normal force contributes to a corresponding static friction force that prevents the person from sliding down the wall to the floor.

Most frequently, the normal force arises on flat horizontal surfaces as a response to the weight of an object pressing it down. As a result, the normal force is often equal to an object's weight, so the forces of friction are equal to the coefficients of friction multiplied by the object's weight. This assumption is false only on non-horizontal surfaces (for example, slopes) or when additional forces are acting to push an object down or hold it up.

TENSION:

The attached object experiences the same pulling force from the opposite end of the wire while a cable pulls an object. Tension is the term for the internal stretch force of a material when force is applied through rope or other pullable materials.

Any force unit, including N, lbs, and other units, can measure tension. The assumption that stress is equal throughout the material is a common simplification. But this presupposes the material's mass is either 0 or extremely small.

Tension is highest where force is being applied. Imagine a 100-pound weight dangling on a 10-pound chain to comprehend why this is the case. The chain's bottom link performs a crucial function by supporting the 100-pound weight. As a result, the chain is under 100 pounds of tension at this point.

However, the chain link just above it also serves a crucial purpose. It must have a tension equal to 100 pounds plus the weight of one chain link because it supports the link supporting the 100-pound weight. So if you keep going up the chain, where the applied force keeps it in place, the top link supports the 100 pounds of weight and an additional 10 pounds of the chain.

The chain's tension at this point is 110 pounds. The lesson here is that the most likely location for a rope or cable to shatter is not close to the thing it is holding or drawing but rather close to where the rope or cable is being pulled, barring flaws or weaknesses in the material under tension. When a heavy-duty cable is drawn in by a winch (a machine used to wind a rope) and the tension exceeds the material's tensile strength, the cable will probably snap extremely near to the winch rather than close to the load.

HYDRAULIC PRESSURE:

Another method for gaining a mechanical advantage is fluid power. The use of fluid to transmit force is known as hydraulics. While hydraulics is a little more adaptable at rerouting and amplifying forces in complex systems, classic basic machines are still valuable and dependable for many applications.

It is first required to have a firm understanding of pressure to comprehend how hydraulics functions. Pressure is applied to a region if a force is distributed uniformly. The formula P =, where F is force in pounds (lb), A is area in square inches (in2), and P is pressure in pounds per square inch (psi), is used to compute pressure.

The following units are more typical in hydraulics, whereas other pressure uses in physics and chemistry utilize newtons and square meters. Do not forget that A divides F. A pressure of 10 psi is created when a force of 100 pounds is applied over a surface area of 10 square inches. You can calculate the force created when pressure P is applied to area A by using the pressure formula to derive F = PA, which is helpful.

A slight volume reduction will occur even when a liquid is subjected to extraordinarily high pressure. Since almost no effort is required to compress liquids into a smaller space, this property of near-incompressibility makes liquids extraordinarily efficient at transmitting force. Furthermore, liquids fill any space in their container if there is enough liquid.

Blaise Pascal, a French mathematician and physicist, created Pascal's law, which states that pressure in any area of an enclosed fluid is constant and points in all directions. So, increasing pressure on one end of a hydraulic system will make it more powerful overall.

When applied to an area of the same size, this increased pressure will produce the same force but in a different place and direction. The location that a force acts on can alter how much force is applied there. For a hydraulic system to deliver a greater force over a shorter distance, a smaller force must be applied over a longer distance.

TORQUE:

The amount that a force can twist an item is measured as torque. Like pressure, torque can't be referred to as a force in and of itself. The formula for torque is r = rF, where r is the length of the lever arm, measured from the pivot point (center of rotation) or fulcrum to the point where the force is applied, and the Greek letter tau stands for torque. Normal units of measurement for force and lever arm length are N or lb and m or ft, respectively. Therefore, torque can be expressed in either the traditional N-m (newton-meters) unit or the ft-lb (foot-pound) unit.

Conversely, Torque is not identical to work, whose N-m units (more commonly written as joules, J) describe a force applied over a distance. Remember that this length measurement is perpendicular to the force used in spinning the object, not parallel. Likewise, torque is not the same as energy, measured in J.

The force applied to rotation and the distance from the center of rotation where the force acts are directly proportional to torque. Applying force at the farthest end of a lever arm is recommended for the highest torque. A longer wrench might be the solution when a mechanic uses all his force to turn a challenging bolt but still has no luck.

While using a longer lever arm for a given force does not increase the amount of force he can apply, it does increase the torque, which should improve his chances of successfully loosening the bolt. Contrary to work, torque is stated to have been exerted before an object has successfully rotated.

Energy, Work, and Power:

Energy is the ability to change the state of anything. Although there are many distinct types of energy, the Mechanical Comprehension section of the ASVAB focuses primarily on mechanical energy. To respond to MC test questions, you must comprehend the two types of mechanical energy—kinetic and potential. Kinetic energy is the force behind the motion.

According to your own experience, a pitch that you might throw at 25 meters per second has less kinetic energy than one that a major league pitcher throws at a speed of 40 meters per second. Similarly, a bicycle moving at the same pace as a car at 10 meters per second has more kinetic energy. Therefore, it would be accurate to state that kinetic energy must depend on mass and velocity.

The formula to calculate kinetic energy (KE) is:

$KE = \frac{1}{2} mv^2$ where m is mass in kilograms (kg), v is speed in meters per second $\left(\frac{m}{s}\right)$, and KE is kinetic energy in a standard unit called joules (J), which has the units $\left(\frac{kg \times m^2}{s^2}\right)$. Note that KE is a linear function of mass, but the velocity component is squared.

As its name suggests, potential energy (PE) can be transformed into kinetic energy. Gravitational potential energy is a common type of PE used in ASVAB questions. An object's potential energy is completely transformed to kinetic energy (KE) when it strikes the ground if you hold it and raise it above it. The object's potential energy increases with the height it is raised to (h). Since heavy and light things fall at the same pace, a heavier object dropped from the same height would have a higher KE when it touched the ground and a higher PE when held above it. Of course, gravity is the force pulling on the dropped object. On the Moon, the pull of gravity is much weaker than on Earth. Thus, dropping an object from the same height would hit the Earth with less force. The object's PE would be lower on the Moon as a result.

The formula to calculate gravitational potential energy follows the logic above.

$PE = mgh$ where m is mass in kilograms (kg), g is the **acceleration due to gravity** $\left(9.8 \frac{m}{s^2}\right)$, h is the height of the object in meters (m), and PE is gravitational potential energy measured in joules (J).

On the ASVAB, it does not matter whether the object is dropped or rolled down a ramp inclined planes#gravitational potential energy because minor factors such as friction will be ignored on energy problems; the velocity, KE, and PE will be the same in either case, even though it will take longer for the object on the ramp to reach the ground.

WORK:

When force is used to move an object, work is completed. The formula for calculating work is consistent with what you can see in the real world, just like the methods for calculating energy. The required work increases with the force to move an object a given distance. Similarly, more work is required to move an object farther when applying force.

The following formula enumerates these details:

$W = Fd$,

Where d is the distance in meters (m), F is the force in newtons (N), and W is the work in joules (J). A car can be moved 10 meters by applying a force of 100 N, resulting in a total work of 100 x 10 = 1,000 joules.

No work has been done if a force is applied to a car, yet the car remains stationary. This is demonstrated by the equation $W = Fd$, where zero distance (d) leads to zero work (W). No movement causes any work to be completed, regardless of the force used. Take note that joules are used to measure both work and kinetic energy. This is not a coincidence because any effort made to move an object from rest to velocity v will result in the conversion of that object's kinetic energy. The work-energy theorem is the name of this concept.

POWER:

Power is the **rate** at which work is done:

$$P = \frac{w}{t}$$

Consider the example of a woman pushing a small car over a distance of 100 meters. If the woman pushes with a force of 100 N to move the car, then using $W = F \times d$, you can see that 100 N × 100 m = 10,000 J of work has been done. If she accomplishes the task at a constant rate over 500 seconds, then the power applied was $\frac{10,000}{500} = 20 \frac{J}{s}$. If another woman can do the same amount of work in 250 seconds, then that woman has twice as much power, or $40 \frac{J}{s}$. One standard unit used to express power is the watt, which equals $1 \frac{J}{s}$ Or $1 \frac{N \times m}{s}$.

The Watt is named after James Watt, an engineer who died in 1819 and whose work on the steam engine helped make the Industrial Revolution possible. The unit of power that many people in the United States are more familiar with is **horsepower** since this is the unit that is used to rate internal combustion engines. Horsepower uses a different system of measurement: it is defined as 550-foot pounds per second of work being done. One horsepower is the equivalent of 746 watts.

Simple Machines:

Simple machines were created ages ago to lessen the force or effort required to carry out various activities, such as lifting heavy objects, bringing together or separating parts, and moving massive equipment. However, a person frequently finds it difficult or impossible to exert enough force to complete the task. Therefore, mechanical advantage is utilized to enhance the available force considerably bigger.

The definition of mechanical advantage is "the advantage gained by the use of a mechanism in transmitting force." The force used can be multiplied by multiple using the right tools. However, even if mechanical advantage can be exploited to increase force, doing so requires applying a force across a longer distance. So, the work output can never be greater than the work input.

The good news is that by dividing complex tasks into smaller, more manageable portions, simple devices can make it easier. How much of the power applied to a machine is converted into motion or force is referred to as its efficiency.

A machine would be 100% efficient if it converted 100% of the inputted power into outputted force. However, no machine is utterly efficient in the actual world since friction affects all machines, and the majority are affected by worn-out and/or improperly fitted parts.

INCLINED PLANE:

Using an inclined plane is among the easiest ways to handle lifting chores. It is difficult, if not impossible, for one person to lift a box weighing 100 kg (approximately 220 pounds). However, an inclined plane (or ramp) can make this task manageable.

Remember that the effort required to lift the box will equal its final gravitational potential energy (PE = mgh). As a result, whether the box is lifted directly or pushed up the ramp (assumed there is no friction), the same amount of work will be required. However, the ramp may raise the box with much less force than the other method. Even though this force must be supplied over a greater distance, it will have some mechanical benefits that will simplify the task.

WEDGE:

The wedge modifies the inclined plane. The wedge is made to move while the inclined plane stays put. It has a variety of uses, such as lifting, splitting, and tightening. The ratio of a wedge's length to height determines the mechanical advantage it can provide.

The lift that occurs as the wedge is moved horizontally will decrease if it is made longer than its height (i.e., with a lesser slope). As a result, the wedge can generate more force, which facilitates lifting big things. Remember that if the wedge increases force, the amount it will lift will be less than the amount it moved horizontally. The wedge is frequently used with knives, chisels, and wood splitters.

LEVERS:

The lever is another straightforward device that may increase force or distance and alter direction. Three basic types of levers make this machine suitable for a wide range of tasks. The first-class lever extends or widens the range of motion and alters the force's direction. A child's teeter-totter is an illustration of a first-class lever.

First-Class Lever:
The lever's pivotal point is known as the fulcrum. The movement at the other end of the lever is in the opposite direction when force is applied to one end. Any mechanical advantage will depend on where the fulcrum is located. If the fulcrum is close to it, less force is needed to lift the object.

To accomplish the task, the end to which the effort is directed must go further. When the fulcrum is placed nearer the location of the effort, the reverse result occurs. However, the thing can now be moved a greater distance with a greater force.

Said the mechanical benefit of using a lever is the ratio of the applied force's distance from the fulcrum to the object's distance from the fulcrum. A lever will, therefore, have a mechanical advantage of 3:1 if the force is exerted 3 meters from the fulcrum and the object being operated upon is 1 meter from the fulcrum.

The force on the object will be three times as strong as the force applied, even if the force applied would propel the object three times as far. A second-class lever exerts more force on the item in the same direction as the applied force. With this kind of lever, lifting a heavier load needs less force, but it must be applied farther. A wheelbarrow is an example of a second-class lever.

Moving the object closer to the fulcrum increases the second-class lever's mechanical advantage. The lever's length may be extended to achieve the same result.

The **third-class lever** also increases the distance the object travels in the same direction as the force applied. It is similar to the second-class lever in that the fulcrum is at one end. The object, however, is at the other end of the lever, with the force being applied somewhere between the fulcrum and the object. Examples of third-class levers would be a fishing pole and a catapult.

Since the distance moved at the point where the force is applied is less than the distance traveled by the load with a third-class lever, the force needed is increased. This effect magnifies as the force is applied closer to the fulcrum.

PULLEYS:

A pulley can occasionally be used to exert force on an object. A pulley system is used to change the direction of a force but not its magnitude.

A pulley system involves a rope, belt, or chain looped over a wheel. It can be useful in certain situations because pulling with a downward force is often easier than an upward force.

BLOCK AND TACKLE:

It is relatively uncommon for one pulley to be used by itself. Often, two or more pulleys will be used in an arrangement known as a **block and tackle** to increase lifting force. The block and tackle shown here have four pulleys; two are attached to a stationary object (such as the ceiling or wall of a building), and the other two are attached to the lower block (the load).

The lower block moves toward the stationary pulleys as the rope is pulled. Note that to lift the load 1 foot, pulling the rope a total of 4 feet is necessary. This is because each of the four rope links must be shortened by 1 foot to get the lower block to move 1 foot.

Neglecting friction gives a total mechanical advantage of 4:1, so if 4 pounds of force are required to lift a load, only 1 pound must be applied to the rope. Note that you can determine the mechanical advantage ratio of a block and tackle system by counting the number of pulleys.

WHEEL AND AXLE:

The wheel and axle mechanism is more complex than a straightforward free-turning wheel like a standard bicycle. Instead, this machine uses two wheels mounted on a shaft, one with a broader diameter than the other. It is possible to increase mechanical advantage by using the wheel and axle. The steering wheel of an automobile is an illustration of a wheel and axle.

The wheel with the largest diameter will often receive the most force—the huge wheel moves in large motions, which causes the little wheel to move in small ones. A point on the edge of the larger wheel rotates farther per turn than a point on the smaller wheel because the larger wheel has a larger circumference.

The force is thus enhanced since a huge movement (or distance) is converted into a little movement. The ratio of the wheel sizes determines how much mechanical advantage there is. The total mechanical advantage is 2:1 if the little wheel is 10 inches in diameter and the giant wheel is 20 inches in diameter.

GEARS AND GEAR RATIOS:

Gears are ideal for a mechanical advantage since they may be used everywhere that force is transferred between two places.

Changes in rotational speed and torque can be made using gears. The gear ratio can be calculated by comparing the number of teeth on the large gear to the number on the small gear. For example, the main gear in the illustration has 18 teeth, while the little gear has 12, making the gear ratio 3:2.

A speed reduction occurs when a tiny gear drives a large gear. The large gear will rotate more slowly than the tiny gear, and the output speed will be proportionately slower as the gear ratio increases. The torque output of the large gear will be more than the torque supplied to the small gear, proportionate to the gear ratio, due to the corresponding torque multiplication (increase). The torque will rise as speed decreases. These two characteristics are said to be inversely proportional by a mathematician.

Mechanical Comprehension Practice Set 1:

1. The work required to lift an object is the same as:
(A) the change in potential energy of the object
(B) the weight of the object divided by gravity
(C) the force required to overcome friction
(D) half of the work required putting the object down

2. A constant force of 2 N is applied to a mass of 6 kg that is initially at rest for 3 seconds. Assuming there is no friction or any other force applied to the mass, what is the velocity of the mass after 3 seconds?
(A) 0.5 m/s
(B) 1.0 m/s
(C) 6.0 m/s
(D) 12.0 m/s

3. Approximately how much force is needed to lift the weight below?

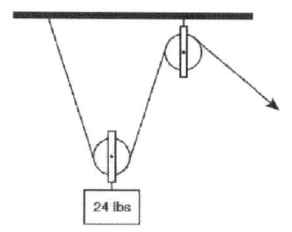

(A) 6 lbs
(B) 12 lbs
(C) 24 lbs
(D) 48 lbs

4. If a ball is dropped from a height of 5 m, what will be its approximate speed when it hits the ground?
(A) cannot be determined because the mass is unknown.
(B) cannot be determined because the time it takes to reach the ground is unknown.
(C) 5 m/s
(D) 10 m/s

5. Adam's car is stuck in the mud, so he recruits some friends to help push it out. How much work did they accomplish if they applied 2,000 N of force to push the car 5 m?
(A) 400 J
(B) 1,000 J
(C) 2,000 J
(D) 10,000 J

6. A machine in an assembly plant lifts parts weighing 10 N each from the floor to a height of 2 m. If this machine must lift batches of 20 parts at a time in 2 seconds, what amount of power is required?
(A) 20 watts
(B) 100 watts
(C) 200 watts
(D) 400 watts

7. A man who weighs 900 N wants to compress some loose soil in his garden, so he puts a thick plate that is m wide and m long on the ground and stands on the plate. Assuming the plate distributes its weight evenly, how much pressure is applied to the soil?
(A) 900 $\frac{N}{m^2}$
(B) 1600 $\frac{N}{m^2}$
(C) 1800 $\frac{N}{m^2}$
(D) 2400 $\frac{N}{m^2}$

8. Gear A has 15 teeth, and Gear B has 10. If gear A makes 14 revolutions, how many will gear B make?

(A) 7
(B) 14
(C) 21
(D) 28

9. Two wheels are secured to a common axle. If the larger wheel has a m radius and the smaller wheel has a m diameter, what is the mechanical advantage of a force applied to the larger wheel?
(A) 5 to 1
(B) 10 to 1
(C) 20 to 1
(D) 40 to 1

10. A basic kitchen knife, like the one below, is based on which of the following simple machines?

(A) wedge
(B) first-class lever
(C) pulley
(D) second-class lever

11. How much force is needed to apply 50 ft-lb of torque to a bolt using a 2 ft-long wrench?
(A) 100 pounds
(B) 50 pounds
(C) 40 pounds
(D) 25 pounds

12. If a 10 N force applied to an object produces an acceleration of 10 m/s2, assuming friction is NOT negligible, what could the object's mass have been?
(A) less than 1 kg

(B) equal to 1 kg
(C) between 10 kg and 1 kg
(D) greater than 10 kg

13. The force of friction between two surfaces moving past each other DOES NOT depend on
(A) the normal force
(B) the nature of the surfaces in contact with each other
(C) the area of the surfaces in contact with each other
(D) the speed at which the surfaces are moving past each other

14. The gravitational acceleration on Earth, g, is 9.8 m/s2. The gravitational acceleration on a neighboring planet, Xendor, is 3 times as much. What is the approximate weight of a 100 kg person on Xendor? (Hint: you can approximate the gravitational acceleration on Earth as 10 m/s2 to make calculations easier.)
(A) 100 N
(B) 300 N
(C) 1,000 N
(D) 3,000 N

15. Assuming no friction, what was the net force applied if 80 J of work was done to move an object 2 m?
(A) 4 N
(B) 16 N
(C) 40 N
(D) 160 N

Mechanical Comprehension Practice Set 2

16. A man pushes against an object at rest but fails to get it moving. Which of the following must be true about the relationship between the applied force and the force of friction?
(A) The applied force is less than the force of static friction.
(B) The applied force is equal to the force of kinetic friction.
(C) The applied force is equal to the force of static friction.
(D) The applied force is greater than the static friction force.

17. A ball with a mass of 10 kg starts rolling down a hill 20 m high. What is its approximate velocity when it hits bottom, ignoring resistance?
(A) 10 m/s
(B) 15 m/s
(C) 20 m/s
(D) 40 m/s

18. A 500 g block made from a uniform material is suspended from the ceiling by two equal-length ropes equidistant from the center of the block, as depicted in the figure below. Which of the following best approximates the tension in rope 1 (T_1)?

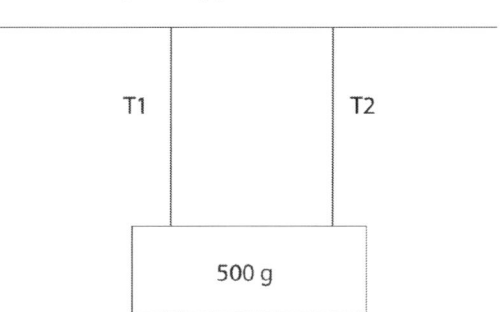

(A) 2.5 N
(B) 5.0 N
(C) 50 N
(D) 250 N

19. A hydraulic lift is set up as depicted below. Which of the following must be true?

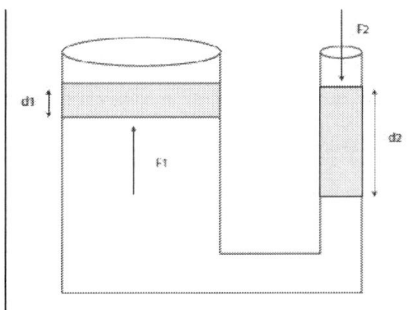

(A) The work done by Force 2 is greater than that done by Force 1 since Force 2 is applied over a longer distance.
(B) To do the same amount of work, Force 2 must be smaller than Force 1 but applied over a larger distance.
(C) Force 1 must be greater than Force 2 because more work must be done by Force 1.
(D) The pressure on the left side of the lift must be greater than the pressure on the right side since Force 1 is larger.

20. Two weights, one with a mass of 10 kg and another with a mass of 20 kg, are placed on opposite ends of a 9 m rod. To achieve rotational equilibrium, what distance must the fulcrum be placed from the 10 kg weight?

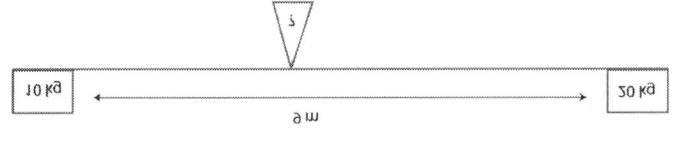

(A) 3.0 m
(B) 4.5 m
(C) 6.0 m
(D) 8.0 m

21. What is the kinetic energy of a 98 N object moving at 100 m/s?
(A) 10,000 J
(B) 50,000 J
(C) 100,000 J
(D) 500,000 J

22. A man pushes a 50 kg block at a distance of 10 meters in 20 seconds with a force of 100 N. How much power did he exert?
(A) 25 W
(B) 50 W
(C) 100 W
(D) 200 W

23. Based on the diagram below, approximately what minimum force must be applied to the rope to lift the object?

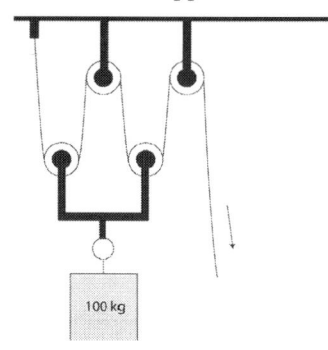

(A) 25 N
(B) 50 N
(C) 100 N
(D) 250 N

24. Snowshoes reduce the pressure exerted by the feet when walking on top of deep snow. A 150 lb woman puts on two rectangular snowshoes with dimensions 10 inches by 15 inches each. What pressure does the woman exert on the snow while both shoes bear her full weight equally?
(A) 0.5 psi
(B) 1.0 psi
(C) 5.0 psi
(D) 10.0 psi

25. A 40 kg block sits on a table, and a 100 N force also pushes it down. What is the magnitude of the table's normal force if the object remains at rest? (Use 10 m/s2 for gravity.)
(A) 140 N
(B) 300 N
(C) 400 N
(D) 500 N

26. A ball with a mass of 5 kg is thrown from a height of 10 meters at an angle of 30° above the horizontal with an initial velocity of 10 m/s. Consider any resistance to be negligible. What is the approximate total kinetic energy when the ball hits the ground?
(A) 250 J
(B) 300 J
(C) 500 J
(D) 750 J

27. Which of the answer choices best describes the topics included in the physics area known as "mechanics?"
(A) the proper use of tools and measuring devices
(B) machines that transmit or use energy
(C) the laws of motion, energy, and forces
(D) the laws of trajectories, energy, and forces

28. Once sufficient force is applied to a stationary object to overcome static friction, the object will accelerate if the same force is maintained. This is true because _____.
(A) the force that is applied accumulates
(B) the coefficient of kinetic friction is greater than the coefficient of static friction
(C) the coefficient of static friction is greater than the coefficient of kinetic friction
(D) the object's momentum is cumulative

29. Playground seesaws are constructed in the same way as first-class levers. A boy wants to play on the seesaw with his little brother. The big brother weighs 100 pounds; the little brother weighs 75 pounds and sits 8 feet from the seesaw's center. How far from the center of the seesaw should the big brother sit for the two brothers to be balanced?
(A) 6.0 feet
(B) 8.0 feet
(C) 10.0 feet
(D) 12.5 feet

30. A hydraulic lift with a 2 cm diameter piston exerts a downward force of 20 N on the incompressible fluid in a closed system. If the piston at the other end has a diameter of 20 cm, how much upward force can it exert?
(A) 20 N
(B) 200 N
(C) 1,000 N
(D) 2,000 N

# Answers and Explanations
MECHANICAL COMPREHENSION PRACTICE SET 1

1. A
The work-energy theorem states that work performed on an object will increase the object's mechanical energy by the amount of work applied to the object. In this case, the increased mechanical energy was the additional potential energy resulting from the object being lifted higher. Therefore, the final *PE* equals the work.

2. B
Since $F = ma$, then $a = F/m = 2/6 = ⅓$ m/s² Change in velocity = acceleration × time. Therefore: ⅓ m/s² × 3s = 1m/s

3. B
This block and tackle has 2 pulleys, so the mechanical advantage is 2:1 because the weight will only move half the distance that the rope is pulled. As a result, the force applied will be half the weight or 12 lbs.

4. D
Although the mass and time are unknown, they are unnecessary to solve the problem. When the ball is 5 m above the ground, all its energy is *PE*, but when it hits the ground, all its energy is *KE*. Since energy is conserved, set $PE = KE$ or $mgh = \frac{1}{2}mv^2$ The mass cancels out, and since the question asks for the approximate speed, $g = 10$ will suffice. 2 × 10 × 5 = v2 and 100 = v2, so v = 10.

5. D
Work = Force × Distance or 2000 × 5 = 10,000 J.

6. C
To calculate power, start by determining the work performed. Since 20 parts weigh (exert a downward force) 10 N each that is lifted 2 m, work = 20 × 10 × 2 = 400 J. The machine accomplishes this work in 2 seconds. Power is $\frac{work}{time}$, or 200 watts.

7. D
Pressure is force divided by the area over which it is applied, so $P = \frac{900}{\left(\frac{1}{2}\right)\times\left(\frac{3}{4}\right)} = \frac{900}{\frac{3}{8}} = 900 \times \frac{8}{3} = 2400 \frac{N}{m^2}$.

8. C
This could be solved with two different approaches. Gear A makes 14 revolutions with 15 teeth, so 14 × 15 = 210 teeth would pass any point. Since gear B has 10 teeth, it would need to make 21 revolutions to match the teeth of gear A. Alternatively, the gear ratio is 3:2, so the proportion $\left(\frac{3}{2}\right) = \left(\frac{x}{14}\right)$ could be used to obtain the same answer, 21

9. B
The mechanical advantage of a wheel and axle is equal to the ratios of the radii of the two wheels. The problem states that the smaller wheel's diameter is $\frac{1}{10}$ m, so its radius is $\frac{1}{20}$ m, and the ratio of the two radii is 10:1.

10. A
The cross-section of a kitchen knife is very small on the cutting side and gets thicker toward the top, just as a wedge does.

11. D
Torque (τ) is determined by multiplying a force (*F*) by the lever arm length (*r*) that the force is acting through or = rF. In this case, achieving an applied torque of 50 ft-lb with a 2 ft lever arm requires F =,/ /25 lb of force.

12. A
Based on Newton's second law, $F = ma$, where F = force in newtons, m = mass in kilograms, and a = acceleration in m/s2. Therefore, by rearranging the equation for mass, you get $m = \frac{F}{a}$. Since you know that friction is not negligible, you also know that the resultant force will be less than 10 N (see diagram below). Therefore, the maximum mass possible in a frictionless system is calculated by dividing 10 N by 10 m/s2, which gives 1 kg. Knowing that friction will diminish the applied force, the resulting mass has to be less than 1 kg, which matches answer choice (A).

10 N applied force   ? N opposing frictional force   Resultant Force = 10 N – the force of friction. Thus, the resultant force will be less than 10 N.

13. D

Kinetic friction is exhibited when surfaces/objects move past one another. The normal force, a factor of the object's weight, will affect friction (for example, move your hand across a table lightly and then again with more force; increased force leads to increased friction). The nature and area of the surfaces in contact with each other will also affect the friction that develops between two surfaces (for example, moving your hand across a desk with oil on it is a lot easier than with honey; also, moving one finger across a desk develops less friction than moving your entire hand). However, the speed at which the surfaces move past each other does not impact the frictional forces (though more heat may be generated).

14. D

Weight = *gm*, where *g* = gravitational acceleration (m/s2) and *m* = mass in kg. On Xendor, the gravitational acceleration is 3 times as much as Earth's, giving a new *g* of approximately 30 m/s2. To calculate the weight on Xendor, multiply the new gravitational acceleration by the person's mass: Weight = 30 m/s2 × 100 kg = 3000 N.

15. C

Recall that *W* = *Fd*, where *W* is work in joules, *F* is force in newtons, and *d* is distance in meters. Therefore, by rearranging the equation for force, you get F= $\frac{W}{d}$. Taking 80 J and dividing by 2 m, you get 40 N.

MECHANICAL COMPREHENSION PRACTICE SET 2

16. C

The object remains at rest, so the acceleration is 0, and the net force must also be 0. Since the object is not moving, the frictional force must be static rather than kinetic. The calculated force of static friction is the maximum that can be attained without creating movement. In this instance, insufficient information is available to determine if this maximum has been reached. However, since the object remains at rest, the applied force must equal the force of static Friction.

17. C

Total mechanical energy is conserved as the ball moves down the hill, meaning that the potential energy at the top will be converted to kinetic energy at the bottom. To solve for the velocity at the bottom, the potential energy must equal kinetic energy. Algebraically, this can be written as $mgh = \frac{1}{2}mv^2$, simplifying to $gh = \frac{1}{2}v^2$ by canceling out mass. Solving for velocity, $v = \sqrt{2gh}$ Finally, plugging in the values from the question stem and using 10m/s2 as an approximation for g, $v = \sqrt{(2)(10)(20)} = \sqrt{400}$ so the ball's velocity when it reaches the bottom of the hill is 20 m/s.

18. A

The question gives the object's mass in grams, so the mass must first be converted to kg by dividing by 1,000, giving 0.5 kg. The mass must then be converted into a force by multiplying by g (use 10 m/s2 for simplicity), so the weight of the mass is 0.5(10) = 5 N. Since the mass is not accelerating, the total tension must be equal and opposite to the object's weight. Algebraically, this means T1 + T2 = 5 N. The tensions are equal because the ropes are of equal length, they are symmetrically attached to the block, and the mass of the block is uniformly distributed. So T1 = T2 and 2T1 = 5 N, so T1 = 2.5 N.

19. B

Since the pressure is the same throughout a closed hydraulic system and force is equal to pressure times the area over which it is applied, F1 must be greater than F2. Also, the work (force times distance) must be equal on both sides. Since F1 > F2, it follows that d2 > d1.

20. C

Since the system is in rotational equilibrium, the net torque on the system must be 0, so the torques exerted by the left- and right-hand sides of the fulcrum must be equal. The question asks the distance from the 10 kg weight; let that distance equal x, so the distance between the fulcrum and the 20 kg weight must be 9 − x. Setting the torques equal each other: 10xg = 20(9 − x)g. Gravity is on both sides of the equation, so it will cancel out, giving 10x = 180 − 20x. Solving for x, x = 6

21. B

The equation for kinetic energy is $\frac{1}{2}mv^2$. Since the object's weight was given, not the mass, the first step is calculating the mass. W = mg, so the equation can be rearranged to $m = \frac{W}{g}$ in order to calculate the mass. Given the weight of 98 N, it will be more accurate to calculate the mass using g = 9.8 m/s2 rather than the approximation of 10 m/s2. So the mass is $\frac{98}{9.8} = 10 \text{ kg}$ Finally, the mass and velocity can be plugged into the equation for kinetic energy: $KE = \frac{1}{2}(10)(100)^2 = 50,000 \text{ J}$

22. B

Since power is work per unit of time and work is force times distance, then $P = \frac{Fd}{t}$ Substituting the known values from the question, $P = \frac{(100)(10)}{20} = 50$ W Note that the mass of the block was not needed to answer this question. Recognizing and discarding extraneous information is a skill that takes practice but is very helpful. 23. **D** The first step is to convert the mass given into a force, which will be its weight. The force of the weight is $mg = 100 \text{ kg} \times 10 \text{ m/s}^2 = 1,000$ N Since there are 4 pulleys, the mechanical advantage is 4:1, meaning that force can be applied over 4 times the distance to pull up the mass. Since the weight is 1,000 N, the force applied would be 250 N.

24. A

Pressure equals force divided by the area over which it is applied. The force is equal to the weight of the woman. To calculate the area, multiply the dimensions of the snowshoes, so $10 \times 15 = 150$ square inches. Since she is wearing two shoes, the total area is $2 \times 150 = 300$ square inches. Divide the force by the area: $\frac{150}{300} = 0.5$ psi

25. D

The normal force must be equal and opposite to the force exerted by the block on the table in accordance with Newton's third law of motion. Since the object's mass is given in kg, the weight must first be calculated using the formula w=mg to get a weight of 400 N. A force is also applied downward on the block of 100 N, so the total force exerted on the table is 500 N. Since the object remains at rest, the normal force must equal 500 N to get a net force of 0 and have no Acceleration.

26. D

The fact that the ball was thrown at an angle may complicate the problem. The ball will travel outward and upward until it reaches its maximum altitude, then travel downward and outward. However, since the question does not specify the direction that the ball is traveling when it hits the ground, the problem can be solved by considering that the total kinetic energy when the ball finally reaches the ground will be the sum of the initial kinetic energy and the potential energy that is converted to kinetic energy. Since air resistance is minimal, no kinetic energy will be lost to air resistance. Therefore, the potential energy will initially be equal to *mgh*. Substituting the known numbers and using the approximation of 10 m/s2 for *g*, you get 5(10)(10) = 500 J. To calculate the beginning kinetic energy, use the equation $KE = \frac{1}{2}mv^2$ Substituting the numbers from the question stem, we Get $\frac{1}{2}(5)(10)^2$ or 250 J. Summing the two forces gives 750 J.

27. C

Tools do use the principles of mechanics, but measuring devices do not. Similarly, machines that transmit or use energy are governed by these principles, but the study of mechanics is not limited to machines. While trajectories are a *type* of motion, the more broad category of "motion" is the proper answer.

28. C

Since the coefficient of static friction is greater than that of kinetic friction, once sufficient force is applied to overcome the maximum value of the force of static friction, and a net force will act to move the object. Because of Newton's second law, *F = ma*, a non-zero net force always creates a non-zero acceleration. That formula applies to the net force, not any force that "accumulates." As the object's velocity increases, so will its momentum, but this is an effect of the acceleration, not the cause.

29. A

In order for the brothers to be balanced, the torque (rotational force) exerted by each boy must be equal. The little brother sits 8 feet from the fulcrum and weighs 75 pounds, so his torque is $8 \text{ ft} \times 75 \text{ lb} = 600$ ft-lb For the big brother to generate 600 ft-lb. Of torque, he should sit 6 feet from the center of the seesaw since $6 \text{ ft} \times 100 \text{ lb} = 600$ ft-lb.

30. D

Since force equals pressure times the area over which the pressure is applied and the pressure is the same throughout the system, the ratios of the forces will be equal to the ratio of the areas of the two pistons. The radius of the smaller piston is 1 cm, and that of the larger piston is 10 cm, so $\frac{\pi(1)^2}{\pi(10)^2} = \frac{20 \text{ N}}{F}$ Cancel the and cross-multiply to get *F* = 2,000 N

CHAPTER 13

Assembling Objects (AO)

Although much of the ASVAB tests academic knowledge at the high school level, Assembling Objects is a subtest that probably doesn't mirror any of your high school classes (unless your high school had a course called Jigsaw Puzzles 101).

The Assembling Objects subtest evaluates your capacity to visualize spatial relationships, analyzing an object's parts to decide how they should fit together. Understanding maps and reading technical drawings requires spatial skills crucial for daily life and academic and professional success. As a result, the need for workers with strong spatial abilities, including the ability to decipher maps, graphs, architectural plans, and X-rays, is rising.

Unless you're taking a test with trial questions, the Assembling Objects subtest of the CAT-ASVAB consists of 15 graphical problems you must resolve in 22 minutes. You have 42 minutes to respond to 30 questions in such a situation. You have 15 minutes to answer 25 questions on the ASVAB paper test. If you're good at jigsaw puzzles, that gives you just under a minute for each question, which is plenty of time to complete.

Understanding the Concept of Object Assembling:

The ASVAB recently added the Assembling Objects subtest. It was included in the 2005 revision of the ASVAB, which also removed the subtests for Coding Speed and Numerical Operations. A year or so later, it was included in the paper enlistment version of the ASVAB after initially only being added to the computerized version. You won't see this subtest if you're taking the ASVAB for high school.

As of this writing, the Navy is the only organization that employs the Assembling Objects subtest score for job qualification. Furthermore, only a few ratings—or professions, as the Navy refers to them—require a score in this area. See Appendix A for information on which Navy enlisted positions call for a particular score.

The results of this subtest aren't used by the other branches right now, but that could change. Therefore, having a high score for this subtest on your resume may help you down the road if you want to change jobs. Otherwise, you might need to retake the test and take this book out of storage. So, read this chapter even if you don't think you need it.

It's a good idea to get it now if you need it to avoid worrying about it later.

Two Types of Questions for the Price of One:

There are two problems on the Assembling Objects subtest, each consisting of five different designs. An image of numerous pieces that have been dismantled is shown in the first drawing, followed by four drawings showing the pieces joined or assembled.

You must select the drawing that best depicts how the components might seem when built or connected correctly. Next, you must mentally spin an array of objects to determine how they would appear if they were turned or rotated a certain number of degrees for both types of Assembling Objects tasks.

Putting Tab A into Slot B: Connectors:

Simple geometric shapes, including stars, clouds, letter shapes, circles, and triangles, are used in the first kind of challenge. You can see shapes and lines in the first drawing marked with dots and the letters A and B. The letters and dots indicate points of attachment here.

The shapes in the following four illustrations are depicted as potential outcomes of how they would appear if a line were to join them at specific locations. From what you can see in the first drawing, the shapes might have been rotated or reoriented. The line is correctly joined in the correct answer to reflect the points in the initial drawing.

See if you can solve Figure 1 by taking a look at it. A star and a somewhat crooked T are depicted in the first illustration. A little dot on the T's short appendage and a dot on one of the star's points, both labeled B, are present.

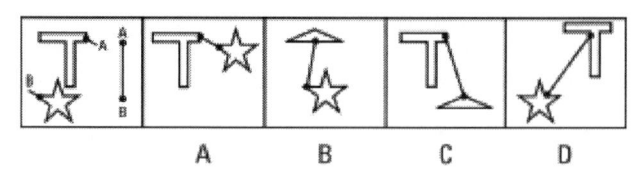

In Figure 1, Choice (A) is the correct solution. Choices (B) and (C) include shapes that aren't included in the first drawing, so they're incorrect. Although Choice (D) has the correct shapes, they aren't connected at the same points depicted in the first drawing.

Okay, that sounds simple. But don't worry; it gets more complicated (sorry to burst your bubble). Figure 2 shows the same problem but with a different twist.

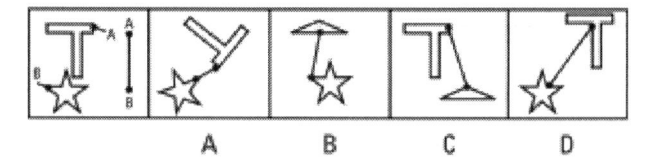

Choice (A) is the correct solution for the problem in Figure 2. In this case, the two shapes have been repositioned and rotated.

Solving the Jigsaw Puzzle: Shapes:

The second category of assembly issues may be perceived by many as being simpler than connection issues. Like a jigsaw puzzle, this kind of challenge doesn't produce a picture of the Statue of Liberty or a map of the United States. Furthermore, these issues have fewer parts than the 1,000-piece puzzle your grandma insisted you assist her with. The challenge comes from the inability to manipulate the tabletop parts with your hands to see how they fit. You have to turn and move the pieces mentally.

In Figure 3, the solution is pretty straightforward.

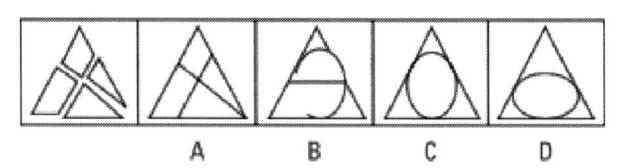

By mentally sliding the shapes in the first drawing together, it's easy to see that they fit together to form the picture shown in Choice (A). Now look at Figure 4.

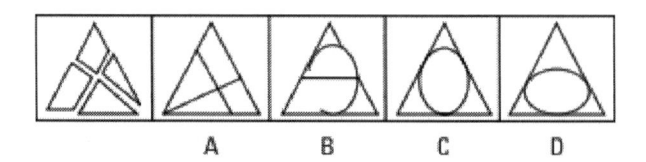

Choice (A) is the correct answer. The figure shown in Choice (A) is the same as that depicted in Choice (A) of Figure 4, except it's been rotated.

The previous two figures were warm-up exercises—the questions on the ASVAB are harder.

Check out Figure 4 for a better representation of the types of questions on the ASVAB.

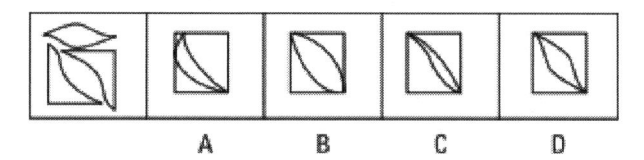

Pay close attention to the leaf-like curvature inside the square. The edges have more of a wave pattern than the bowed-out appearance of Choice (B). Though overly thin, option (C) has that shape. Give yourself a pat if you chose Option (D) as the correct response. Finding the right solution is made simple by looking at spatial relationships. Try a few additional examples to see if you've gotten the feel of it. Take a look at Figure 4.

Choice (B) in Figure 4 is the appropriate response. To visualize how the elements in the first image join together to create the shape in Choice (B), mentally rotate and reposition them. Three pieces in the puzzle have one straight edge and one curved edge. Eliminate options that don't meet these criteria by practicing. Choice (A) can be immediately discounted because it lacks these shapes. The curved parts are of different sizes, as you can see. Compare the sizes and variations of the puzzle pieces as you picture them fitting together. Try Figure 5 now.

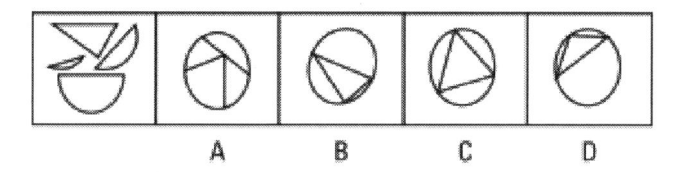

In Figure 5, Choice (A) is the correct answer. If you didn't get this one quite right, don't worry. You can hone your skills with the practice questions at the end of this chapter.

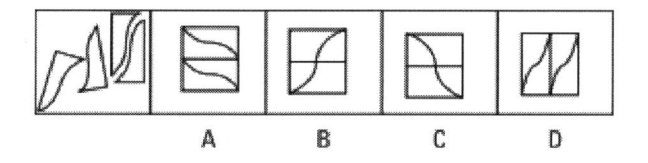

Tips for the Assembling Objects Subtest:

In the following sections, I advise raising your Assembling Objects subtest score. I offer tips for minimizing incorrect responses on the exam and ways to enhance your overall spatial abilities (which may be helpful the next time you have to read a map).

Comparing one piece or point at a time:

When taking the Assembling Objects subtest, picking just one form from the first drawing and rapidly scanning the choices to see if it is present but in a different orientation might occasionally increase your chances of getting the appropriate answer. You can quickly eliminate answer options that are blatantly incorrect using this technique.

On connection-type problems, note where the dot is in the first drawing on one of the shapes, and then quickly scan the answers, eliminating any option that shows the dot or line passing through the shape somewhere other than what is shown in the first drawing.

Be mindful of mirror images, which are shapes that are reflected (as opposed to rotated) from the image depicted in the initial drawing. Cunning test writers frequently use mirror images to determine if they can deceive your sight.

Practicing Spatial Skills:

Researchers at the University of Chicago have determined that your basic foundation for spatial skills is established at a very early age, perhaps as young as age 4 or 5. But don't worry. That doesn't mean all is lost if your parents never got you that model rocket kit you wanted. The same research has concluded that you can still improve spatial skills by engaging in spatially oriented activities.

Some of those activities include the following:

» **Practicing Reading Maps**: Map reading can help you develop the ability to gauge scales of size and direction between related objects (roads, rivers, towns, cities, and so on).
» **Putting Together Jigsaw Puzzles**: This way is an obvious form of practice for improving your spatial perceptions.
» **Playing Puzzle Games Online**: Many online puzzle games exercise the skill of identifying spatial relationships and visual similarities.
» **Playing Graphical Computer Games**: Computer games may help to improve your spatial skills. A study conducted in the United Kingdom showed that children who played computer games consistently scored higher on spatial aptitude tests than children who didn't play the Games.
» **Sketching**: Look at an object or a picture and attempt to sketch it as viewed differently. This exercise can help to improve your ability to visualize angles mentally.

Assembling Objects Practice Questions:

Assembling Objects questions measure your spatial abilities. Connecting questions and putting pieces together questions are the two different categories of inquiries. You are responsible for selecting the response in connection questions that best depict how the shapes are joined at the specified spots.

You must select the response in the jigsaw puzzle-style questions that most accurately depicts what the initial drawing's shapes would be like when put together.

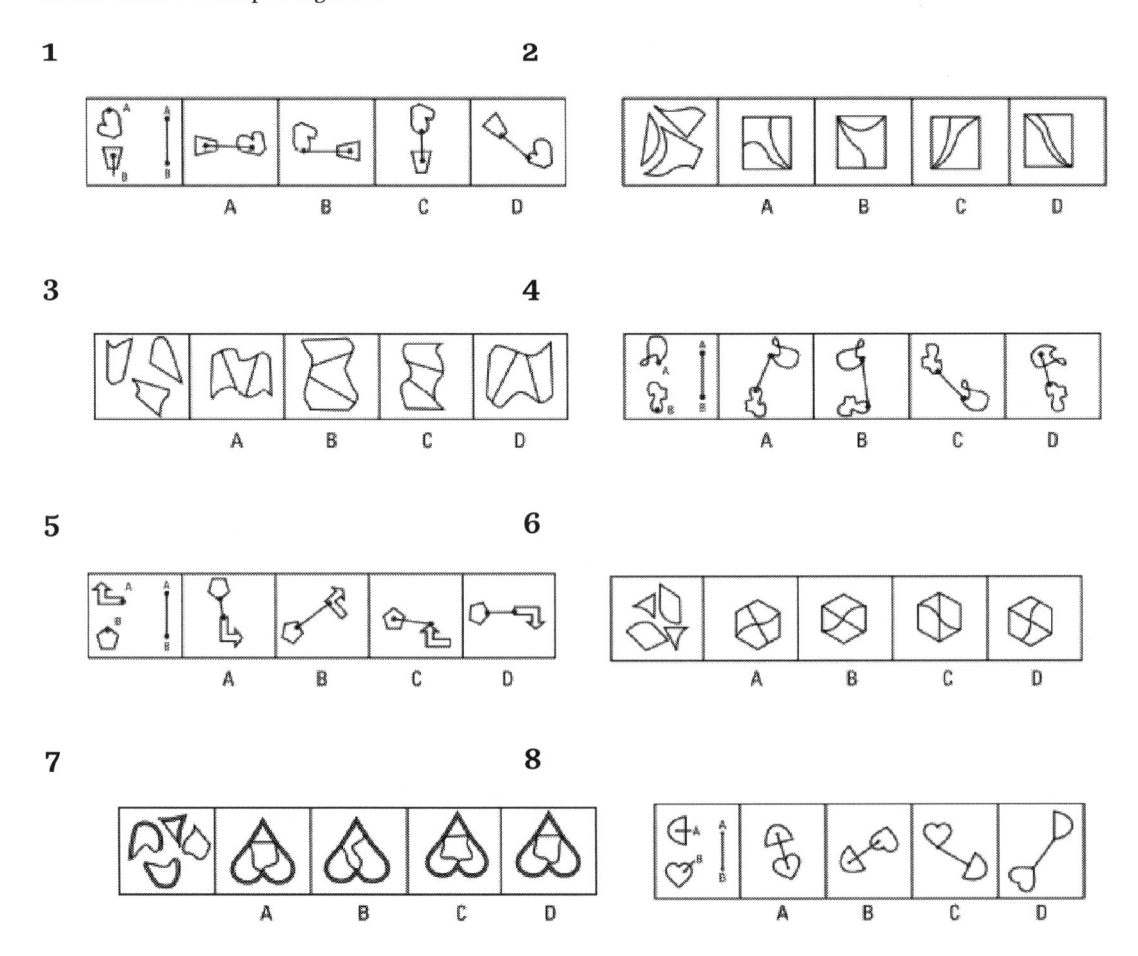

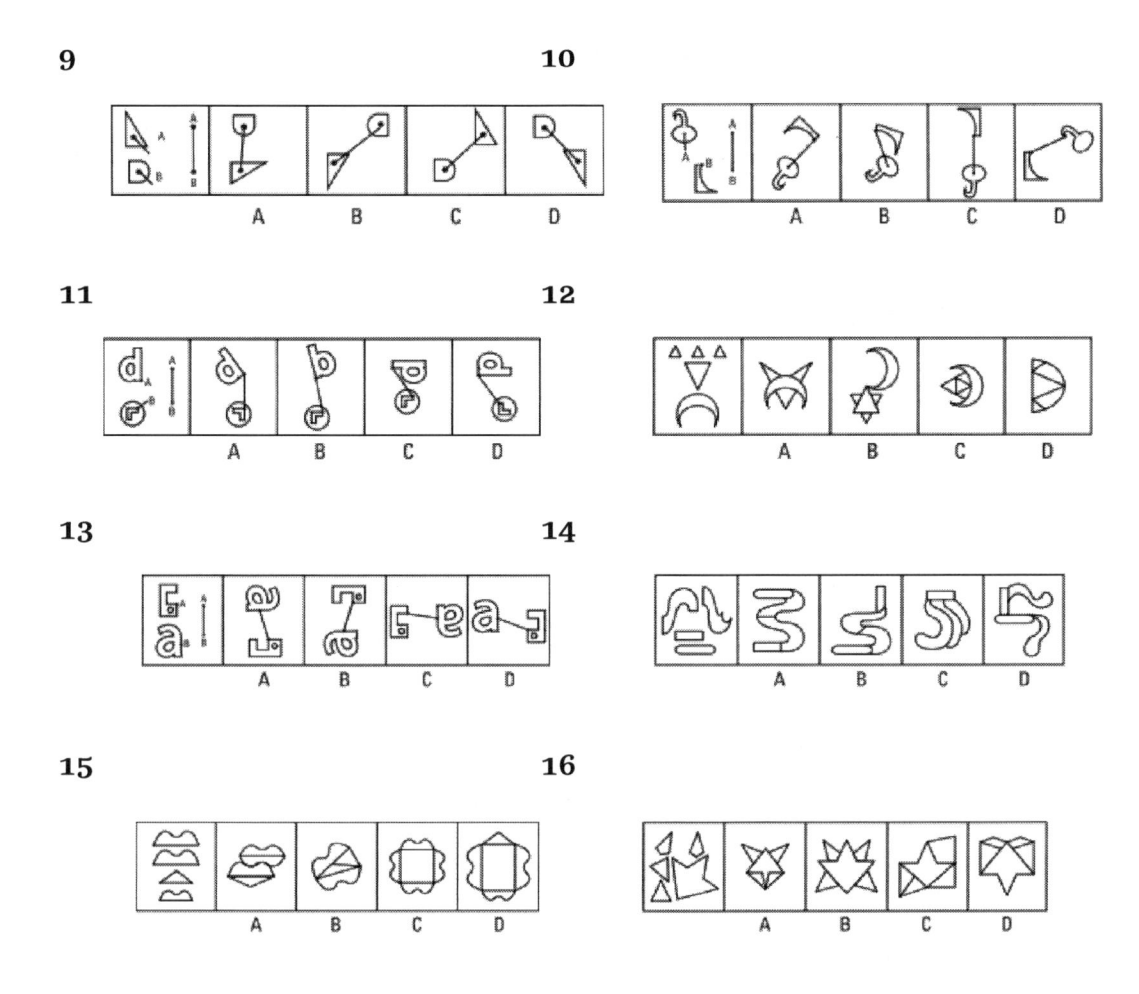

Answers and Explanations:

Use this answer key to score the Assembling Objects practice questions.

1. B.
Note that the bottom figure in the first drawing has a line that intersects the short side of the trapezoid shape, so Choices (C) and (D) are wrong. Connection point A is at the tip of the mitten shape, so Choice (A) is also wrong. The correct answer is Choice (B).

2. C.
Mentally rotate and reposition the shapes in the first drawing until you can see how they fit together to form the shape shown in Choice (C)—the correct answer. In the first drawing, notice that the shape at the upper right resembles a shark fin—it has two sharp points, and the third point is curved. Choice (C) is the only image that contains this shape (it's at the bottom).

3. D.
If you selected Choice (A), you were fooled. The arrow shape in Choice (A) mirrors the shape depicted in the first drawing. The correct answer is Choice (D).

4. A.
Mentally rotate and reposition the shapes in the first drawing until you can see how they fit together to form the shape shown in Choice (A)—the correct answer. If you had trouble with this one, notice that the piece in the center of the upside-down heart should have a corner that dips a bit on the left. Choice (C) has the dip in the center, and Choice (D) is on the right, so these answers are wrong. Choice (B) has only three pieces.

5. D.
Note that both shapes in the first drawing have lines that intersect the shapes at designated points. If you selected Choice (B), mirror images fooled your eyes. The correct answer is Choice (D).

6. B.

Mentally rotate and reposition the shapes in the first drawing until you can see how they fit together to form the shape shown in Choice (B)—the correct answer. Here, you can take a mental snapshot of the largest shape and look for it in the answers—Choice (B) is the only choice. Verify that this is the right answer by recognizing that Choice (B) is the only answer containing a circle segment at the top.

7. A.

Don't be fooled by the mirror shape in Choice (B) because Choice (A) is the correct answer.

8. C.

Mentally rotate and reposition the shapes in the first drawing until you can see how they fit together to form the shape shown in Choice (C), which is the correct answer. Here, you may note that the first drawing contains two shapes that resemble triangles, with one side curved inward. Choice (C) is the only image that contains those shapes.

9. B.

Take note of the point of intersection in the heart of the question, then match it up with the correct answer, Choice (B).

10. A.

Awkward shapes plus mirrored images make this one a little tricky, but when you look at the points and the positions of the images, you can see that Choice (A) is correct.

11. D.

Choice (D) is the only answer that shows the right intersection between the circle and the *L*-shaped object.

12. A.

Keeping your eye on the points of assembly and staying clear of mirrored images, you can see that Choice (A) is connected appropriately.

13. D.

The rectangle in the middle of Choices (C) and (D) is a negative space. However, you can tell Choice (D) is correct by sizing the proportions of the three curvy shapes and the triangle.

14. B.

Counting out the shapes and identifying the right proportions help you see Choice (B) as your shining star.

15. A.

Awkward shapes can make it challenging to piece together multiple objects with the correct proportions mentally. Don't be distracted by the shapes. Notice the proportions and lines in relation to one another. Make sure each element in the question appears in your answer.

16. B.

Unusual shapes can be awkward to dissect, but Choice (B) reflects the correct assembly. It has the right number of shapes in the right proportions.

CHAPTER 14

Practice Exam 1:

General Science:

1. Which planet is named after the Greek god who personified the sky?
(A) Earth
(B) Mars
(C) Pluto
(D) Uranus

2. An animal that eats only meat is called a(n):
(A) omnivore.
(B) herbivore.
(C) carnivore.
(D) voracious.

3. The chemical process in which electrons are removed from a molecule is called:
(A) respiration.
(B) recreation.
(C) oxidation.
(D) metabolism.

4. What is a single unit of quanta called?
(A) quantum
(B) quantumonium
(C) quantus
(D) quanfactorial

5. In a vacuum, light waves travel at a rate of about:
(A) 186,000 miles per hour.
(B) 186,000 miles per minute.
(C) 18,600 miles per hour.
(D) 186,000 miles per second.

6. The largest planet in the solar system is:
(A) Earth.
(B) Mars.
(C) Saturn.
(D) Jupiter.

7. The intestines are part of the:
(A) circulatory system.
(B) nervous system.
(C) respiratory system.
(D) digestive system.

8. Joints that hold bones firmly together are called:
(A) hinge joints.
(B) ball and socket joints.
(C) fixed joints.
(D) pivot joints.

9. Of the levels listed, the top or broadest level of the classification system for living organisms is called the:
(A) class.
(B) phylum.
(C) kingdom.
(D) genus.

10. Which planet is the brightest object in the sky besides the sun and moon?
(A) Saturn
(B) Pluto
(C) Venus
(D) Mercury

11. The human heart includes:
(A) 2 chambers.
(B) 3 chambers.
(C) 4 chambers.
(D) 5 chambers.

12. White blood cells:
(A) produce antibodies.
(B) fight infections.
(C) carry oxygen and carbon dioxide.
(D) Both A and B.

13. A measurable amount of protein can be found in all of the following foods EXCEPT:
(A) eggs.
(B) meat.
(C) peas.
(D) apples.

14. What is the most abundant element, by mass, in Earth's crust?
(A) carbon
(B) oxygen
(C) gold
(D) salt

15. Osmosis is:
(A) diffusion of a solvent.
(B) transfer of oxygen.
(C) low blood sugar.
(D) protein.

16. A meter consists of:
(A) 10 centimeters.
(B) 100 millimeters.
(C) 100 centimeters.
(D) 10 millimeters.

17. One light-year is:
(A) the distance traveled by light in one year.
(B) the brightness of light at 30,000 miles.
(C) 17 standard Earth years.
(D) the distance Earth must travel around the sun.

18. Electrons are particles that are:
(A) positively charged.
(B) neutral.
(C) able to move freely.
(D) negatively charged.

19. The asteroid belt is located:
(A) around Mercury.
(B) between Mars and Jupiter.
(C) inside the orbit of Venus.
(D) There is no such thing as an asteroid belt.

20. The atomic number of an atom is determined by:
(A) the size of its nucleus.
(B) the number of protons.
(C) the number of electrons.
(D) its location in the periodic table.

21. The "control center" of a cell is called the:
(A) nucleus.
(B) compound.
(C) mitochondria.
(D) atom.

22. How many planets in the solar system have rings?
(A) one
(B) two
(C) three
(D) four

23. The temperature at which a substance's solid and liquid states exist in equilibrium is its:
(A) melting point.
(B) boiling point.
(C) anti-freezing point.
(D) concentration point.

24. The atmosphere of Mars is composed mainly of:
(A) oxygen.
(B) carbon dioxide.
(C) helium.
(D) Mars has no atmosphere.

25. Not counting the sun, the closest star to Earth is
(A) Rigel.
(B) Proxima Centauri.
(C) Antares.
(D) Betelgeuse.

Arithmetic Reasoning:

1. If a car is towed 12 miles to the repair shop and the tow charge is $3.50 per mile, how much does the tow cost?
(A) $12.00
(B) $3.50
(C) $42.00
(D) $100.00

2. The sum of two numbers is 70. One number is 8 more than the other. What's the smaller number?
(A) 31
(B) 33
(C) 35
(D) 36

3. A sales manager buys antacids in bottles by the gross. If they go through 3 bottles of antacid every day, how long will the gross last?
(A) 144 days
(B) 3 days
(C) 20 days
(D) 48 days

4. Jenny's test grades are 93, 89, 96, and 98. What must she score on her next test if she wishes to raise her average to 95?
(A) 100
(B) 99
(C) 97
(D) 95

5. A server earns an average tip of 12% of the cost of the food they serve. If they serve $375 worth of food in one evening, how much money in tips will they earn on average?
(A) $37
(B) $45
(C) $42
(D) $420

6. How many square feet of carpeting is needed to carpet a 12-foot-x-12-foot room?
(A) 24

(B) 120
(C) 48
(D) 144

7. Carpet stain protector costs $0.65 per square yard to apply. How much will it cost to apply the protector to a 16-foot-x-18-foot carpet?
(A) $187.20
(B) $62.40
(C) $20.80
(D) $96.00

8. A printing plant that produces baseball cards has a monthly overhead of $6,000. It costs 18 cents to print each card, and the cards sell for 30 cents each. How many cards must the printing plant sell monthly to make a profit?

(A) 30,000
(B) 40,000
(C) 50,000
(D) 60,000

9. Joe received an hourly wage of $8.15. His boss gave him a 7% raise. How much does Joe make per hour now?
(A) $0.57
(B) $8.90
(C) $8.72
(D) $13.85

10. Alice leaves her house, driving east at 45 miles per hour (mph). Thirty minutes later, her husband, Dave, notices she forgot her cell phone and sets off after her. How fast must Dave travel to catch up with Alice 3 hours after she leaves?
(A) 49 mph
(B) 50.5 mph
(C) 52.5 mph
(D) 54 mph

11. A baker made 20 pies. A Boy Scout troop buys one-fourth of the pies, a preschool teacher buys one-third, and a caterer buys one-sixth. How many pies does the baker have left?
(A) ¾
(B) 15
(C) 12
(D) 5

12. Miriam bought five cases of motor oil on sale. A case of motor oil typically costs $24.00, but she was able to purchase the oil for $22.50 a case. How much money did Miriam save on her entire purchase?
(A) $7.50
(B) $1.50
(C) $8.00
(D) $22.50

13. A security guard walks the equivalent of six city blocks when they make a circuit around the building. If they walk at a pace of eight city blocks every 30 minutes, how long will it take them to complete a circuit around the building, assuming they don't run into any thieves?
(A) 20.00 minutes
(B) 3.75 minutes
(C) 22.50 minutes
(D) 7.5 minutes

14. The population of Grand Island, Nebraska, grew by 600,000 people between 1995 and 2005, one-fifth more than the town council predicted. The town council originally predicted the city's population would grow by
(A) 400,000
(B) 500,000
(C) 300,000
(D) 100,000

15. Joan is taking an admissions examination. If she has to get at least 40 of the 60 questions right to pass, what percent of the questions does she need to answer correctly?
(A) 30%
(B) 40%
(C) 66 ⅓ %
(D) 66 ⅔ %

16. A teacher deposits $3,000 in a retirement fund. If they don't add any more money to the fund, which earns an annual interest rate of 6%, how much money will they have in 1 year?
(A) $180
(B) $3,006
(C) $3,180
(D) $6,000

17. The high school track measures one-quarter of a mile around. How many laps would you have to run to run three and a half miles?
(A) 12
(B) 14
(C) 16
(D) 18

18. Karl is driving in Austria, where the speed limit is posted in kilometers per hour. The car's speedometer shows that he's traveling at 75 kilometers per hour. Karl knows that a kilometer is about ⅝ of a mile. Approximately how many miles per hour is Karl traveling?
(A) 47
(B) 120
(C) 50
(D) 53

19. A carpenter earns $12.30 an hour for a 40-hour week. Their overtime pay is 1½ times their base pay. If they put in a 46-hour week, how much is their weekly pay?
(A) $602.70
(B) $492.00
(C) $574.66
(D) $110.70

20. An office building has 30 employees and provides 42 square feet of workspace per employee. If five more employees are hired, how much less workspace will each employee have?
(A) 6 square feet
(B) 7 square feet
(C) 7.5 square feet
(D) 36 square feet

21. Stan bought a monster truck for $2,000 down and payments of $450 a month for five years. What's the total cost of the monster truck?
(A) $4,250
(B) $29,000
(C) $27,000
(D) $34,400

22. Darla spent $120.37 on groceries in January, $108.45 in February, and $114.86 in March. What was the average monthly cost of Darla's groceries?
(A) $343.68
(B) $110.45
(C) $114.86
(D) $114.56

23. Keith is driving from Reno to Kansas City to meet his girlfriend. The distance between the two cities is 1,650 miles. If Keith can average 50 miles per hour, how many hours will it take him to complete his trip?
(A) 8 hours
(B) 30 hours
(C) 33 hours
(D) 82 hours

24. Michael needs 55 gallons of paint to paint an apartment building. He would like to purchase the paint for the least amount of money possible. Which of the following should he buy?
(A) two 25-gallon buckets at $550 each
(B) eleven 5-gallon buckets at $108 each
(C) six 10-gallon buckets at $215 each
(D) fifty-five 1-gallon buckets at $23 each

25. As a member of FEMA, you must set up a contingency plan to supply meals to residents of a town devastated by a tornado. A breakfast ration weighs 12 ounces, and the lunch and dinner rations weigh 18 ounces each. Assuming a food truck can carry 3 tons and that each resident will receive 3 meals per day, how many residents can you feed from one truck during a 10-day period?
(A) 150 residents
(B) 200 residents
(C) 250 residents
(D) 300 residents

26. A train headed south for Wichita left the station at the same time a train headed north for Des Moines left the same station. The train headed for Wichita traveled at 55 miles per hour. The train headed for Des Moines traveled at 70 miles per hour. How many miles apart are the trains at the end of 3 hours?
(A) 210 miles
(B) 165 miles
(C) 125 miles
(D) 375 miles

27. A carpenter needs to cut four sections, each 3 feet, 8 inches long, from a piece of molding. If the board is only sold by the foot, what's the shortest length of board the carpenter can buy?
(A) 15 feet
(B) 14 feet
(C) 16 feet
(D) 12 feet

28. Kiya had only one coupon for 10% off one frozen turkey breast. The turkey breasts cost $8.50 each, and Kiya wanted to buy two. How much did she pay?
(A) $16.15
(B) $17.00
(C) $15.30
(D) $7.65

29. A recruiter travels 1,100 miles during a 40-hour workweek. If they ⅖ of their time traveling, how many hours do they spend traveling?
(A) 22
(B) 5 12
(C) 16
(D) 8

30. Your car uses gasoline at 21 miles per gallon. If gasoline costs $2.82 per gallon and you drive for 7 hours at a speed of 48 miles per hour, how much will you pay for gasoline for the trip?
(A) $38.18
(B) $45.12
(C) $47.73
(D) $59.27

Word Knowledge:

1. Tim promised to meet us at the apex.
(A) top
(B) bottom
(C) canyon
(D) river

2. Assimilate most nearly means:
(A) absorb.
(B) react.
(C) pretend.
(D) lie.

3. Brittle most nearly means:
(A) soft.
(B) fragile.
(C) study.
(D) hard.

4. Datum most nearly means:
(A) fiscal year date.
(B) congruence.
(C) fact.
(D) positive result.

5. The exchange student was proficient in French, German, and English.
(A) poor
(B) knowledgeable
(C) adept
(D) exacting

6. The judge imposed a severe penalty due to Tom's actions.
(A) scheduled
(B) made an example of
(C) levied
(D) questioned

7. Mary went to the store and bought peanuts galore.
(A) abundant
(B) salty
(C) on sale
(D) roasted

8. I ran headlong into the fight.
(A) headfirst
(B) reluctantly
(C) happily
(D) recklessly

9. Frugal most nearly means:
(A) quiet.
(B) amazing.
(C) delayed.
(D) economical.

10. The word most opposite in meaning to stimulate is:
(A) support.
(B) arrest.
(C) travel.
(D) dislike.

11. Illicit most nearly means:
(A) historical.
(B) unlawful.
(C) storied.
(D) willfully.

12. Vacate most nearly means:
(A) crawl.
(B) impel.
(C) exhume.
(D) leave.

13. The sergeant gave their reasoned opinion.
(A) irate
(B) logical
(C) impressive
(D) uninformed

14. Tacit most nearly means:
(A) loud.
(B) understood.
(C) commendable.
(D) transparent.

15. The brass was not burnished.
(A) yellow
(B) dull
(C) expensive
(D) polished

16. The commodity was sold.
(A) product
(B) stock
(C) idea
(D) table

17. Your motives were contrived.
(A) premeditated
(B) emotional
(C) obscure
(D) amusing

18. Supplicate most nearly means:
(A) to make superior.
(B) to be unnecessary.
(C) to beg.
(D) to be expansive.

19. The word most opposite in meaning to hypocrisy is
(A) honesty.
(B) happy.
(C) angry.
(D) threatening.

20. Bob found the peaches to be highly succulent.
(A) large
(B) tasteless
(C) old
(D) juicy

21. The Army soldiers were ordered to immediate garrison duty.
(A) field
(B) combat
(C) latrine
(D) fort

22. Furtherance most nearly means:
(A) advancement.
(B) finance.
(C) practicality.
(D) destruction.

23. Domicile most nearly means:
(A) office.
(B) shopping.
(C) home.
(D) vacation.

24. Abrogate most nearly means:
(A) recover.
(B) aid.
(C) foreclose.
(D) abolish.

25. Compensation most nearly means:
(A) religion.
(B) commission.
(C) boathouse.
(D) shower.

26. They gave a brusque account of the events.
(A) passionate
(B) lengthy
(C) uncensored
(D) abrupt

27. The vote resulted in the demise of the proposed new law.
(A) passage
(B) death
(C) postponement
(D) abatement

28. We commemorated our veterans during the ceremony.
(A) denied
(B) remembered
(C) thanked
(D) took pictures of

29. Bore most nearly means:
(A) deepen.
(B) hide.
(C) dig.
(D) jump.

30. That custom still prevails.
(A) angers
(B) persists
(C) surprises
(D) excites

31. Defray most nearly means:
(A) invade.
(B) obstruct.
(C) pay.
(D) reverse.

32. Chasm most nearly means:
(A) abyss.
(B) sky.
(C) mountain.
(D) valley.

33. Fundamental most nearly means:
(A) radical.
(B) religious.
(C) basic.
(D) excessive.

34. Susceptible most nearly means:
(A) travel.
(B) resistant.
(C) limited.
(D) vulnerable.

35. Emblem most nearly means:
(A) symbol.
(B) picture.
(C) statue.
(D) religion.

Paragraph Comprehension:

A critical stage of personal time management is to take control of appointments. Determined by external obligation, appointments constitute interaction with other people and an agreed-on interface between your activities and those of others. Start with a simple appointment diary. List all appointments, including regular and recurring ones. Now, be ruthless and eliminate the unnecessary. There may be committees where you can't productively contribute or where a subordinate may be able to participate. Eliminate the waste of your time.

1. Effectively managing your appointments allows you to:

(A) spend more time with your subordinates.
(B) delegate responsibility to subordinates.
(C) make more efficient use of your time.
(D) attend only the most important Meetings.

The U.S. Congress consists of 100 senators and 435 representatives. Two senators are elected from each state. The number of representatives from each state is based on population, although each state has at least one representative. Senators serve six-year terms, and representatives serve two-year terms.

2. According to this passage,

(A) there are equal numbers of senators and representatives.
(B) the number of representatives from each state is decided by a lottery.
(C) it's possible for a state to have no representatives.
(D) senators and representatives have different term lengths.

Indo-European languages consist of those languages spoken by most of Europe and in those parts of the world that Europeans have colonized since the 16th century (such as the United States). Indo-European languages are also spoken in India, Iran, parts of western Afghanistan, and in some areas of Asia.

3. The author of this passage would agree that:

(A) Indo-European languages are spoken in areas all over the world.
(B) Indo-European languages include all the languages spoken in the world.
(C) only Europeans speak Indo-European languages.
(D) Indo-European language speakers can easily understand one another.

In privatization, the government relies on the private sector to provide a service. However, the government divests itself of the entire process, including all assets. With privatized functions, the government may specify quality, quantity, and timeliness requirements, but it has no control over the operations of the activity. Also, the government may not be the only customer. Whomever the government chooses to provide the services would likely provide the same services to others.

4. This paragraph best supports the statement that:

(A) the government must closely supervise privatized functions.
(B) privatized functions consist of a mixture of government employees, military personnel, and private contractors.
(C) privatized functions are those institutions that provide services only to a government agency.
(D) privatized functions provide essential services to the government.

The success or failure of a conference lies largely with its leader. A leader's zest and enthusiasm must be real, apparent, and contagious. The leader is responsible for getting the ball rolling and making the attendees feel the meeting is theirs and its success depends on their participation. A good, thorough introduction helps establish the right climate.

5. A good title to this paragraph would be:

(A) "Lead by Example."
(B) "The Importance of Proper Introductions."
(C) "Leading a Successful Conference."
(D) "Conference Participation Basics."

Cloud seeding is accomplished by dropping particles of dry ice (solid carbon dioxide) from a plane onto super-cooled clouds. This process encourages condensation of water droplets in the clouds, which usually, but not always, results in rain or snow.

6. From this passage, it's reasonable to assume that:

(A) cloud seeding could be used to end a drought.
(B) cloud seeding is prohibitively expensive.
(C) cloud seeding is rarely used.
(D) cloud seeding can be accomplished by using regular ice.

To write or not to write—that is the question. If assigned a writing task, there's no option.

However, if someone is looking for a specific answer, find out if they need a short or detailed answer. Can the requirement be met with a telephone call, email, or short note, or is something more necessary? A former CEO of a major corporation once commented that he had looked at 13,000 pieces of paper in 5 days. Think how much easier and more economical it would be if people would use the telephone, send an email, or write a short note.

7. The main point of this passage is that:

(A) written records are essential because they provide detailed documentation.
(B) more businesspeople should invest time and energy in improving their writing skills.
(C) writing may not be the best way to communicate information.
(D) it's pointless for businesspeople to spend time improving their writing skills.

The transistor, a small, solid-state device that can amplify sound, was invented in 1947. At first, it was too expensive and too difficult to produce to be used in cheap, mass-market products. By 1954, though, these cost and production problems had been overcome, and the first transistor radio was put on the market.

8. According to this passage,

(A) there was no market for transistors before 1954.
(B) the transistor radio was put on the market when transistors could be produced cheaply and easily.
(C) transistors were invented in 1947 by order of the Department of Defense.
(D) transistors are still expensive to produce.

I returned from the City about three o'clock on that May afternoon, disgusted with life.

I had been three months in the Old Country and was fed up with it. If people had told me a year ago that I would've been feeling like that I should've laughed at them; but there was the fact. The weather made me liverish, the talk of the ordinary Englishman made me sick, I couldn't get enough exercise, and the amusements of London seemed as flat as soda water that had been standing in the sun.

9. The author is speaking of his travels in:

(A) Spain.
(B) Great Britain.
(C) Germany.
(D) Scotland.

Surveys show that the average child under the age of 18 watches four hours of television per day. Although some of the programming may be educational, most isn't. Spending this much time watching television interferes with a child's ability to pursue other interests, such as reading, participating in sports, and playing with friends.

10. The author of this passage would agree that:

(A) television viewing should be restricted.
(B) parents who let their children watch this much television are neglectful.
(C) reading, participating in sports, playing with friends, and watching television should all be given equal time.
(D) adults over 18 can watch as much television as they want.

Questions 11 and 12 are based on the following passage.
High school and college graduates attempting to find jobs should participate in mock job interviews. These mock interviews help students prepare for the types of questions they'll be asked, make them more comfortable with common interview formats, and help them critique their performance before facing a real interviewer. Because they're such a valuable aid, schools should organize mock job interviews for all their graduating students.

11. The above passage states that mock job interviews:

(A) frighten students.
(B) should be offered to the best students.
(C) help prepare students for real job interviews.
(D) should be organized by students.

12. From the above passage, it is reasonable to assume that:

(A) mock interviews can increase a student's confidence when they go into a real job interview.
(B) mock interviews are expensive to organize.
(C) few students are interested in mock interviews.
(D) students don't need job interview Preparation.

Questions 13 through 15 are based on the following passage.
Due process, the guarantee of fairness in the administration of justice, is part of the 5th Amend ment to the U.S. Constitution. The 14th Amendment further requires states to abide by due process. After this amendment was enacted, the U.S. Supreme Court struck down many state laws that infringed on the civil rights guaranteed to citizens in the Bill of Rights.

13. According to the above passage, due process:

(A) is an outdated concept.
(B) guarantees fairness in the justice system.
(C) never became part of the U.S. Constitution.
(D) is the process by which winning lottery tickets are selected.

14. According to the above passage, it's reasonable to assume that the 5th Amendment:

(A) is about taxes.
(B) guarantees due process in all criminal and civil cases.
(C) guarantees due process in federal law.
(D) should never have become part of the Bill of Rights.

15. The author of the above passage would agree that:

(A) without the passage of the 14th Amendment, many laws restricting civil rights would still exist in various states.
(B) the Supreme Court overstepped its jurisdiction when it struck down laws infringing on citizens' civil rights.
(C) the Supreme Court had every right to strike down state laws before the passage of the 14th Amendment.
(D) the 14th Amendment was opposed by all states.

Mathematics Knowledge:

If a train travels at a speed of 60 miles per hour for 2.5 hours, how many miles will it travel?
A. 120 miles
B. 150 miles
C. 200 miles
D. 250 miles

What is the value of x in the equation 3x + 4 = 22?
A. 2
B. 4
C. 6
D. 8

If 5x - 3 = 27, what is the value of x?
A. 5
B. 6
C. 7
D. 8

A car is sold for $25,000, 80% of its original price. What was the original price of the car?
A. $20,000
B. $31,250
C. $32,500
D. $40,000

The length of a rectangle is twice its width. If the perimeter of the rectangle is 36 inches, what is the length of the rectangle?
A. 6 inches
B. 8 inches
C. 10 inches
D. 12 inches

What is the sum of the interior angles of a hexagon?
A. 540 degrees
B. 720 degrees
C. 900 degrees
D. 1080 degrees

If a number is decreased by 20% of itself, what is the new number in terms of the original number x?
A. 0.2x
B. 0.8x
C. 0.9x
D. 1.2x

What is the volume of a cube with side length of 5 cm?
A. 25 cubic cm
B. 75 cubic cm
C. 100 cubic cm
D. 125 cubic cm

If the mean of four numbers is 10, what is their sum?
A. 20
B. 30
C. 40
D. 50

What is the slope of the line described by the equation 3y = 6x + 12?
A. 2
B. 4
C. 6
D. 8

Solve for x: 2x - 3 = 4x + 1.
A. -1
B. -2
C. 1
D. 2

What is the value of 3^4?
A. 81
B. 64
C. 48
D. 36

If a tank can be filled in 6 hours and emptied in 9 hours, how long will it take to fill the tank if both the inlet and outlet are open?
A. 18 hours
B. 15 hours
C. 12 hours
D. 9 hours

What is the square root of 256?
A. 12
B. 14
C. 16
D. 18

Solve for y: 4y - 7 = 2y + 5.
A. 3
B. 4
C. 5
D. 6

A car travels 400 miles on 20 gallons of gas. How many miles per gallon does the car get?
A. 10 mpg
B. 20 mpg
C. 25 mpg
D. 30 mpg

If a rectangle has an area of 100 square units and a length of 10 units, what is its width?
A. 5 units
B. 10 units
C. 15 units
D. 20 units

What is the result of 5% of 200?
A. 5
B. 10
C. 20
D. 30

What was the original price if a shirt is sold for $20 after a 20% discount?
A. $15
B. $25
C. $30
D. $40

What is the least common multiple of 6, 8, and 10?
A. 24
B. 30
C. 40
D. 120

If the ratio of cats to dogs in a room is 3:2 and there are 20 animals, how many dogs are there?
A. 6
B. 8
C. 12
D. 14

What is the sum of the first 5 terms of the arithmetic sequence 2, 5, 8, 11?
A. 25
B. 30
C. 35
D. 40

What is the value of x in the equation 2x + 3 = 7?
A. 1
B. 2
C. 3
D. 4

If a number is increased by 15%, what is the new number in terms of the original number x?
A. 0.85x
B. 1.15x
C. 1.25x
D. 1.5x

A bike is sold for $300, 75% of its original price. What was the original price of the bike?
A. $200
B. $400
C. $450
D. $600

What is the sum of the interior angles of a pentagon?
A. 540 degrees
B. 450 degrees
C. 360 degrees
D. 270 degrees

If a = 2 and b = 3, what is the value of $a^2 + b^2$?
A. 7
B. 13
C. 17
D. 25

If you roll a fair six-sided die, what is the probability of rolling a number greater than 4?
A. 1/6
B. 1/3
C. 1/2
D. 2/3

Solve for y: 3y + 5 = 2y - 3.
A. -8
B. -2
C. 2
D. 8

30 What is the value of x in the equation 5x - 2 = 3x + 4?
A. 1
B. 2
C. 3
D. 4

Electronics Information:

1. Ohm's law states:
(A) Voltage Current Resistance
(B) Amperes Current Resistance
(C) Voltage Resistance Amperes
(D) Ohms Current Voltage

2. A resistor's first three color bands are brown, black, and red. What is its value?
(A) 1,000 ohms
(B) 500 ohms
(C) 500 volts
(D) 50 volts

3. In the U.S., all metal equipment, electrical or not, connected to a swimming pool must be:
(A) freestanding.
(B) bonded together.
(C) certified.
(D) none of the above.

4. Voltage can also be expressed as:
(A) watts.
(B) amps.
(C) current.
(D) electrical potential difference.

5. Newer cell phones contain a removable memory card, which is often called a:
(A) SIM card.
(B) DIM chip.
(C) PIN card.
(D) PIN chip.

6. Made from a variety of materials, such as car bon, this inhibits the flow of current.
(A) resistor
(B) diode
(C) transformer
(D) generator

7. This is a type of semiconductor that only allows current to flow in one direction. It is usually used to rectify AC signals (conversion to DC).
(A) capacitor
(B) inductor
(C) diode
(D) transformer

8. Radar can operate at frequencies as high as:
(A) 100,000 Hz.
(B) 100,000 kHz.
(C) 100,000 MHz.
(D) 500,000 MHz.

9. What do AC and DC stand for in the electrical field?
(A) amplified capacity and differential capacity
(B) alternating current and direct current
(C) accelerated climate and deduced climate
(D) none of the above

10. Changing AC to DC is called what?
(A) capacitance
(B) impedance
(C) rectification
(D) induction

11. A 5,000 BTU air conditioner can efficiently cool up to 150 square feet or a 10-foot-x-15- foot room. What does BTU stand for?
(A) basic thermal unit
(B) basic temperature unit
(C) British thermal unit
(D) none of the above

12. Which is the most correct definition of current?
(A) the measure of electrical pressure
(B) the amount of electricity used in a heater
(C) the electricity used in heating a kilo of water
(D) the presence of electron flow

13. A device that transforms energy from one form to another is called:
(A) a capacitor.
(B) a transducer.
(C) a transformer.
(D) magic.

14. Which one of the following is an active element?
(A) 15 k resistor
(B) 10 mH inductor
(C) 25 pF capacitor
(D) 10 V power supply

15. A light bulb is 60 watts. Operated at 120 volts, how much current does it draw?
(A) 0.5 amperes
(B) 5.0 amperes
(C) 50.0 amperes
(D) 7,200 amperes

16. A number-12 wire, compared to a number-6 wire,
(A) is longer.
(B) is shorter.
(C) is smaller in diameter.
(D) is larger in diameter.

17. A fuse with a higher-than-required rating used in an electrical circuit:
(A) improves safety.
(B) increases maintenance.
(C) may not work properly.
(D) is less expensive.

18. Neutral wire is always:
(A) whitish or natural.
(B) black.
(C) green with stripes.
(D) blue.

19. To measure electrical power, you would use a(n):
(A) ammeter.
(B) ohmmeter.
(C) voltmeter.
(D) wattmeter.

20. What will happen if you operate an incandescent light bulb at less than its rated voltage?
(A) The bulb will burn brighter and last longer.
(B) The bulb will burn dimmer and last longer.
(C) The bulb will burn brighter but won't last as long.
(D) The bulb will burn dimmer but won't last as long

Auto & Shop Information:

1. Overheating the engine can cause all of the following problems EXCEPT:
(A) burned engine bearings.
(B) enlarged pistons.
(C) melted engine parts.
(D) improved fuel efficiency.

2. The device that converts an automobile's mechanical energy to electrical energy is called the:
(A) converter.
(B) alternator.
(C) battery.
(D) brakes.

3. A primary advantage of the electronic ignition system over conventional ignition systems is that:
(A) the electronic ignition system is less expensive to repair.
(B) the electronic ignition system requires a lower voltage to provide a higher voltage for the spark.
(C) the electronic ignition system allows for the use of a lower-octane fuel.
(D) all of the above.

4. The primary purpose of piston rings is to:
(A) seal the combustion chamber and allow the pistons to move freely.
(B) connect the piston to the crankshaft.
(C) allow fuel to enter the piston cylinder.
(D) provide lubrication to the piston Cylinder.

5. The crankshaft typically connects to a:
(A) flywheel.
(B) fuel pump.
(C) muffler.
(D) battery.

6. What component allows the left and right wheels to turn at different speeds when cornering?
(A) differential
(B) camshaft
(C) valve rotator
(D) battery

7. If a car's ignition system, lights, and radio don't work, the part that's probably malfunctioned is the:
(A) cylinder block.
(B) water pump.
(C) carburetor.
(D) battery.

8. A good tool to cut intricate shapes in wood would be a:
(A) ripsaw.
(B) hacksaw.
(C) coping saw.
(D) pocket knife.

9. A two-stroke engine will normally be found on:
(A) small cars.
(B) large diesel trucks.
(C) trucks, vans, and some cars.
(D) snowmobiles, chainsaws, and some Motorcycles.

10. A belt sander would best be used to:
(A) cut wood.
(B) finish wood.
(C) shape wood.
(D) keep your pants up.

11. A car equipped with limited-slip differential:
(A) can be readily put into all-wheel (four wheel) drive.
(B) won't lock up when the brakes are applied steadily.
(C) transfers the most driving force to the wheel with the greatest amount of traction.
(D) is rated for off-road driving.

12. Big block engines generally have:
(A) more than 5.9 L of displacement.
(B) better gas mileage than small block engines.
(C) less than 6 L of displacement.
(D) air conditioning.

13. A good tool for spreading and/or shaping mortar would be a:
(A) cement shaper.
(B) hammer.
(C) trowel.
(D) broom.

14. Plumb bobs are used to:
(A) clean pipes.
(B) check vertical reference.
(C) fix the toilet.
(D) carve stones.

15. Rebar is used to:
(A) measure the depth of concrete.
(B) reinforce concrete.
(C) stir concrete.
(D) smooth concrete.

16. Annular ring, clout, and spring head are types of:
(A) hammers.
(B) saws.
(C) nails.
(D) screwdrivers.

17. A ripsaw cuts:
(A) against the grain of the wood.
(B) with the grain of the wood.
(C) most materials, including metal.
(D) only plastic.

18. A cam belt is also known as a:
(A) piston.
(B) timing belt.
(C) transmission belt.
(D) lug nut.

19. To check for horizontal trueness, the best tool to use is a:
(A) steel tape rule.
(B) plumb bob.
(C) level.
(D) sliding T-bevel.

20. A bucking bar is used to:
(A) pull nails.
(B) pry wood apart.
(C) form rivet bucktails.
(D) drive screws.

21. Washers that have teeth all around the circumference to prevent them from slipping are called:
(A) shake-proof washers.
(B) jaw washers.
(C) flat washers.
(D) split lock washers.

Mechanical Comprehension:

1. An induction clutch works by:
(A) magnetism.
(B) pneumatics.
(C) hydraulics.
(D) friction.

2. If a first-class lever with a resistance arm measuring 2 feet and an effort arm measuring 8 feet are being used, what's the mechanical advantage?
(A) 2
(B) 4
(C) 6
(D) 1

3. The bottoms of four boxes are shown below. The boxes all have the same volume. If postal regulations state that the sides of a box must meet a minimum height, which box is most likely too short to go through the mail?

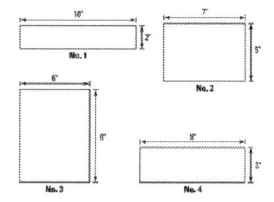

(A) No. 1
(B) No. 2
(C) No. 3
(D) No. 4

4. Looking at the figure below where Anvil A and B have the same mass when Anvil B lands on the seesaw, Anvil A will:

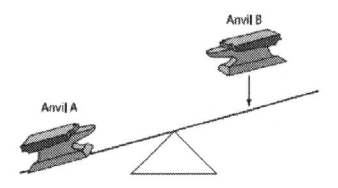

(A) remain stationary.
(B) hit the ground hard.
(C) rise in the air quickly.
(D) enter the stratosphere.

5. Air pressure at sea level is about 15 psi. What's the amount of force exerted on the top of your head, given a surface area of 24 square inches?
(A) 360 pounds
(B) 625 pounds
(C) ⅝ pound
(D) 180 pounds

6. The force produced when a boxer's hand hits a heavy bag and "bounces" off it is called
(A) response time.
(B) bounce.
(C) recoil.
(D) gravity.

7. In the figure below, if Gear 1 has 25 teeth and Gear 2 has 15 teeth, how many revolutions does Gear 2 make for every 10 revolutions Gear 1 makes?

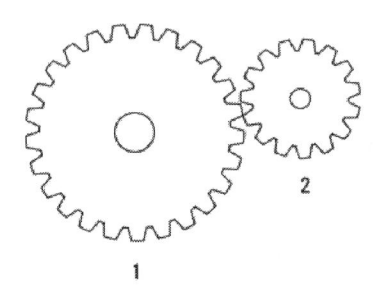

(A) about 16⅔
(B) 12
(C) about 13 more
(D) about 20

8. A cubic foot of water weighs about 62.5 pounds. If an aquarium is 18 feet long, 10 feet deep, and 12 feet wide, what's the approximate pressure in pounds per square inch (psi) on the bottom of the tank?
(A) 2 psi
(B) 4 psi
(C) 5 psi
(D) 7 psi

9. Springs used in machines are usually made of:
(A) plastic.
(B) bronze.
(C) nylon fiber.
(D) steel.

10. A clutch is a type of:
(A) universal joint.
(B) coupling.
(C) gear differential.
(D) cam follower.

11. When Cam A completes one revolution, the lever will touch the contact point:

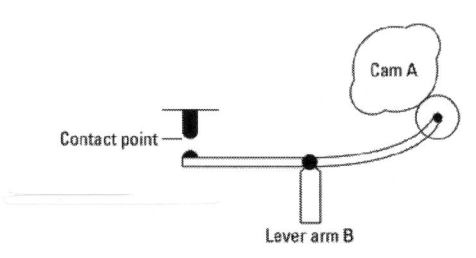

(A) once.
(B) never.
(C) four times.
(D) twice.

12. A single moveable block-and-fall is called a:
(A) fixed pulley.
(B) gun tackle.
(C) runner.
(D) sheave.

13. In the figure below, if the fulcrum supporting the lever is moved closer to the anvil, the anvil will be:

(A) easier to lift and will move higher.
(B) harder to lift but will move higher.
(C) easier to lift but will not move as high.
(D) harder to lift and will not move as high.

14. The mechanical advantage of the block and-tackle arrangement shown below is

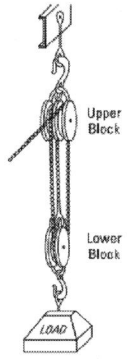

(A) 2
(B) 3
(C) 6
(D) 1

15. In the figure below, if the cogs move up the track at the same rate of speed, Cog A will:

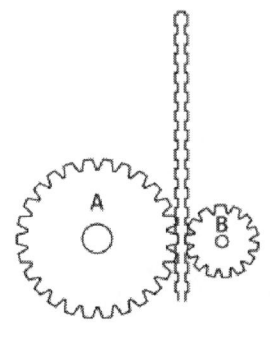

(A) reach the top at the same time as Cog B.

(B) reach the top after Cog B.
(C) reach the top before Cog B.
(D) have greater difficulty staying on track.

16. If a house key, a wooden spoon, a plastic hanger, and a wool jacket are all the same temperature on a cool day, which one feels the coldest?
(A) key
(B) spoon
(C) hanger
(D) jacket

17. In the figure below, assume the valves are all closed. To fill the tank but prevent it from filling entirely, which valves should be open?

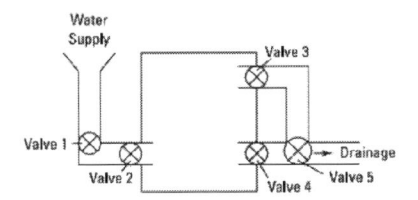

(A) 1 and 2 only
(B) 1, 2, and 3 only
(C) 1, 2, and 4 only
(D) 1, 2, 3, and 5 only

18. If Gear A is turned to the left,

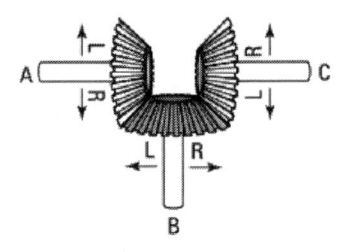

(A) Gear B turns to the right and Gear C turns to the left.
(B) Gear B turns to the left, and Gear C turns to the left.
(C) Gear B turns to the right and Gear C turns to the right.
(D) Gear B turns to the left and Gear C turns to the right.

19. If Gear 1 moves clockwise, which other gears also turn clockwise?

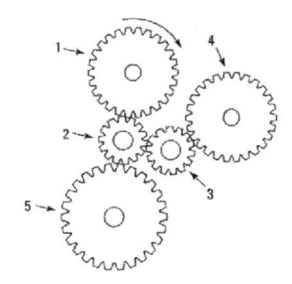

(A) 3 and 5
(B) 3, 4, and 5
(C) 2 and 5
(D) 3 and 4

20. The pressure gauge in the figure below shows a reading of:

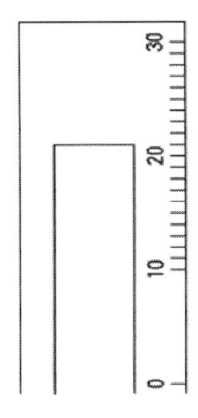

(A) 15.0
(B) 19.5
(C) 21.0
(D) 23.0

21. A way to determine the amount of power being used is to:
(A) multiply the amount of work done by the time it takes.
(B) multiply the distance covered by the time it takes to move a load.
(C) divide the amount of work done by 550 pounds per second.
(D) divide the amount of work done by the amount of time it takes.

22. A wood tool, a silver tool, and a steel tool are placed in boiling water for cleaning. Which tool will get hot most quickly?
(A) steel
(B) wood
(C) silver
(D) All three are equally hot

23. A whip is being used in the figure shown. How much effort is the boy who's lifting the 50-pound anvil using? Disregard friction, wind resistance, and the weight of the pulley and the rope.

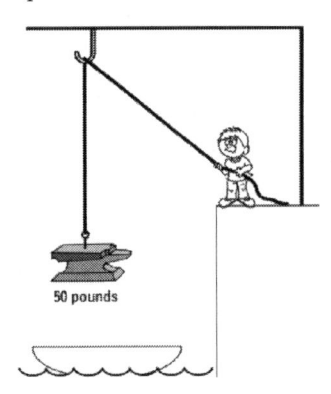

(A) 50-pound effort
(B) 100-pound effort
(C) 25-pound effort
(D) 10-pound effort

Assembling Object:

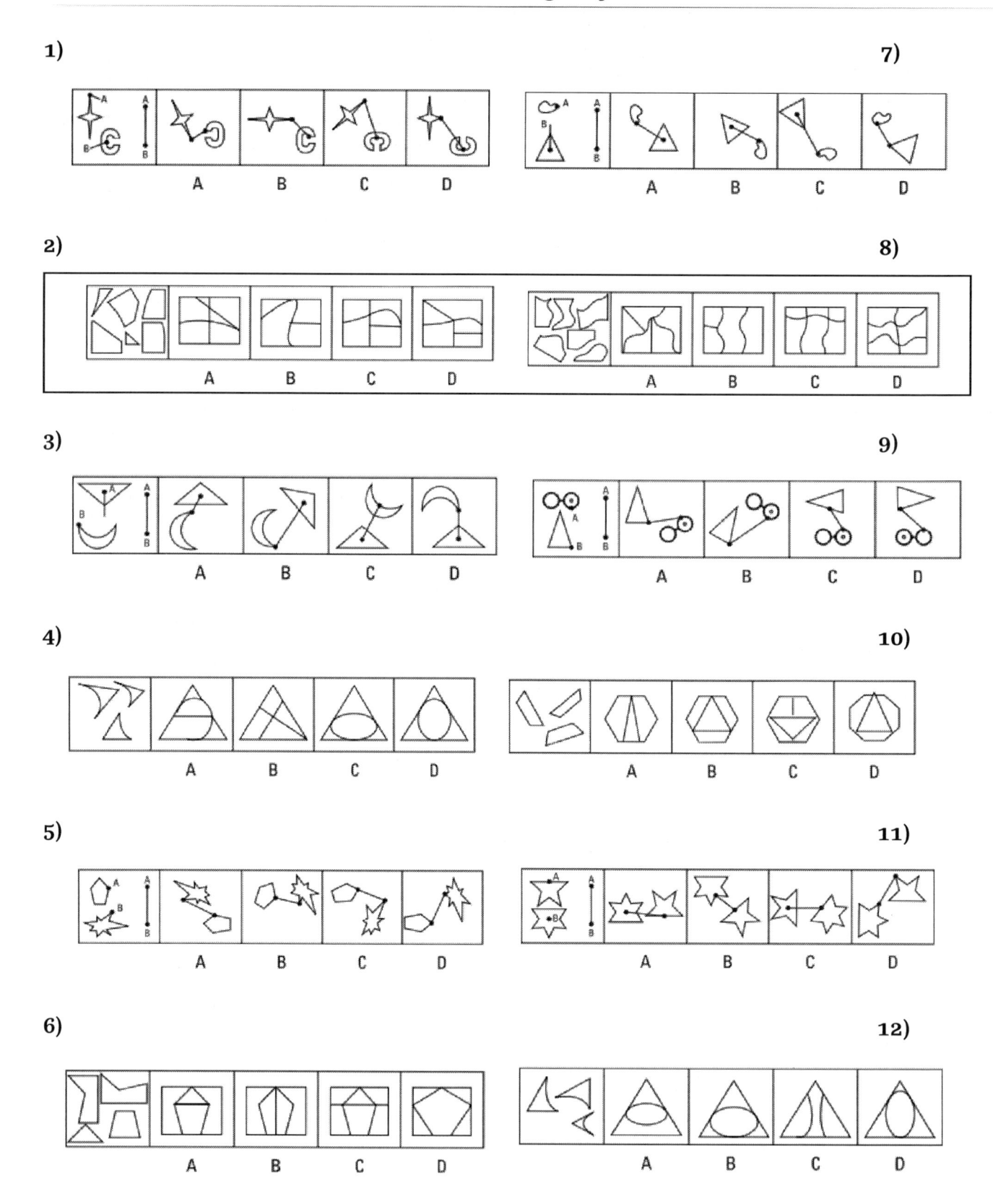

1)

7)

2)

8)

3)

9)

4)

10)

5)

11)

6)

12)

13)
14)
15)
16)
17)
18)
19)
20)
21)
22)
23)
24)
25)

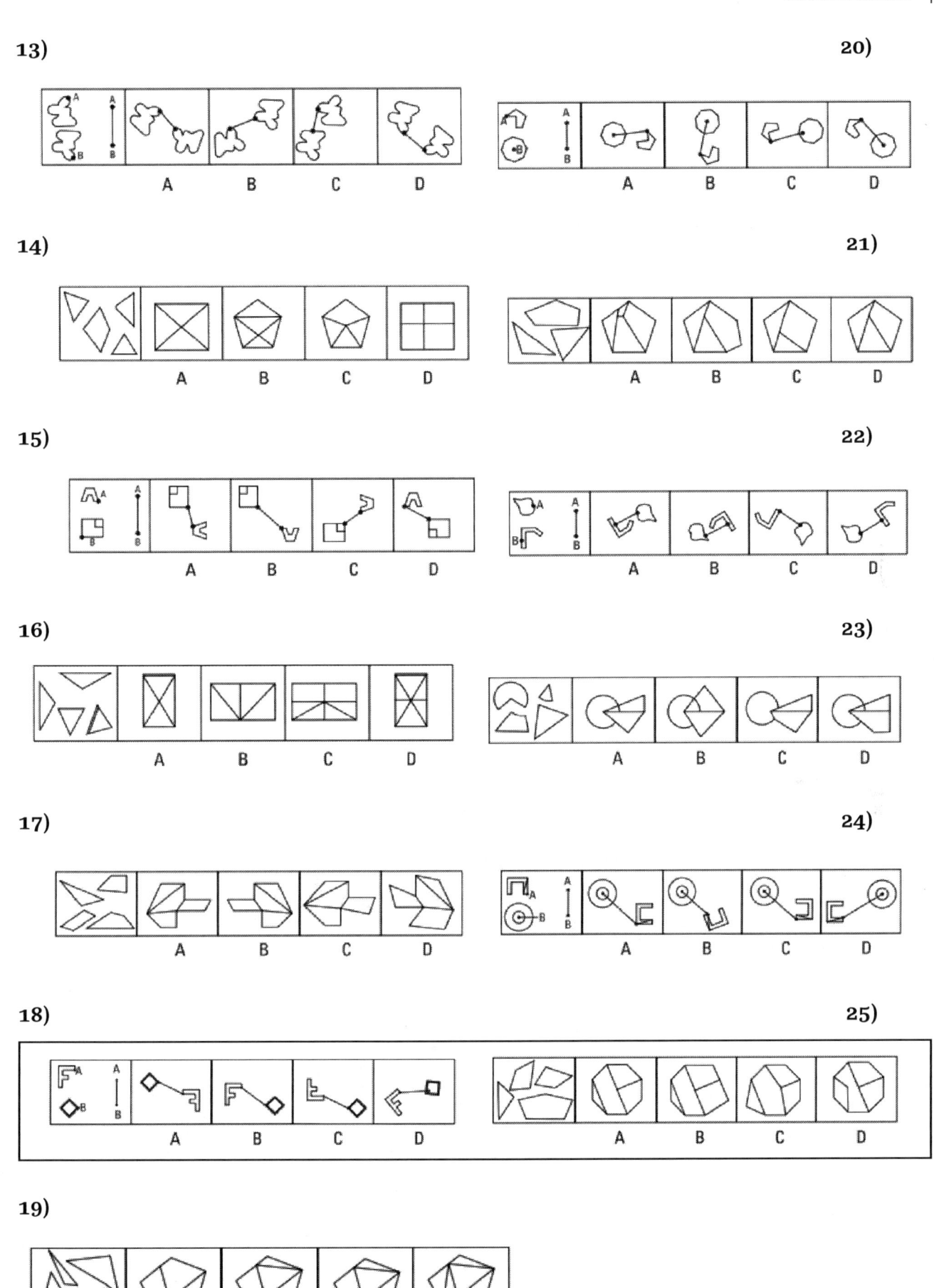

Practice Exam 1 Answers and Explanations

General Science Answers

1. D.
The planet Uranus was named for the Greek god Uranus, who personified the sky.

2. C.
From the Latin, *carne* means flesh, and *vorare* means to devour—hence the word *carnivore*, a meat-eater. Therefore, choice (C) is correct.

3. C.
In the chemical process of *oxidation,* an atom loses electrons in its *valence shell* (the outermost shell of electrons). Oxidation is usually paired with *reduction,* in which another atom or molecule gains those electrons. Together, reduction and oxidation form a *redox* reaction.

4. A.
The plural of *quantum* is *quanta*.

5. D.
Light waves travel at 299,792 kilometers (or approximately 186,000 miles) per second. That's why light from the sun takes about 8.3 minutes to reach Earth.

6. D.
Jupiter is the largest planet in our solar system.

7. D.
Both the large and small intestines are part of the digestive system.

8. C.
Joints that hold bones firmly together are called *fixed joints*.

9. C.
The levels of classification of living creatures from broadest to narrowest are kingdom, phylum, class, order, family, genus, and species.

10. C.
Venus is the brightest planet and is sometimes called the *morning* or *evening star.*

11. C.
The left and right ventricles, with the left and right atria, make up the four heart chambers in a human.

12. D.
White blood cells produce antibodies and fight infection, making Choice (D) correct.

13. D.
Eggs, meat, and peas all contain protein. However, a negligible amount of protein occurs in apples.

14. B.
Oxygen makes up about 46.6 percent of the Earth's crust. Silicon is the next-most abundant, coming in at 27.7 percent.

15. A.
Osmosis is a passive process in which water or another solvent moves through a partially permeable membrane. If the solution on one side of the membrane has a high concentration of solute and the other has a lower concentration of solute, the water moves through the membrane toward the highly concentrated side until the concentrations are equal.

16. C.
A centimeter is one-hundredth of a meter; therefore, you find 100 centimeters in a meter.

17. A.
The distance light travels in one year is one *light-year*.

18. D.
Electrons are the negatively charged parts of an atom.

19. B.
The asteroid belt is located between the planets Mars and Jupiter.

20. B.
The number of protons in an atom's nucleus determines the atomic number of an atom.

21. A.
A cell's nucleus contains its nuclear genome. It's sometimes called the control center.

22. D.
Jupiter, Saturn, Neptune, and Uranus are the four planets with rings.

23. A.
The melting point of a substance is a state of equilibrium between its solid and liquid states. The melting point is equal to the freezing point.

24. B.
Mars's atmosphere is comprised mainly of carbon dioxide.

25. B.
Proxima Centauri C (or Alpha Centauri C) is the closest star to the Earth (after our sun).

Arithmetic Reasoning Answers

1. **(C) $42.00**
Explanation: 12 miles * $3.50 per mile = $42.00.

2. **(A) 31**
Explanation: Let x be the smaller number, then the other is x + 8. x + (x + 8) = 70. Solving, x = 33.

3. **(D) 48 days**
Explanation: A gross is 144 bottles, so 144 bottles / 3 bottles per day = 48 days.

4. **(B) 99**
Explanation: Average of first four grades = (93 + 89 + 96 + 98) / 4 = 94. She needs a 99 on the next test for a 95 average.

5. **(B) $45**
Explanation: 12% of $375 = 0.12 * $375 = $45.

6. **(D) 144**
Explanation: 12-foot * 12-foot = 144 square feet.

7. **(C) $20,80**
Explanation: 16 feet * 18 feet = 288 square feet = 32 square yards. 32 * $0.65 = $62.40.

8. **(C) 50,000**
Explanation: Profit per card = $0.30 - $0.18 = $0.12. Profit needed = $6000. $6000 / $0.12 = 50,000 cards.

9. **(B) $8.90**
Explanation: 7% raise of $8.15 = 0.07 * $8.15 = $0.57. $8.15 + $0.57 = $8.90.

10. **(D) 54 mph**
Explanation: Alice has a 22.5 miles head start (30 minutes * 45 mph). Dave must make up 22.5 miles in 3 hours and drive 7.5 mph faster than Alice. 45 mph + 7.5 mph = 54 mph.

11. **(D) 5**
Explanation: Boy Scouts: 20 * 1/4 = 5 pies, preschool teacher: 20 * 1/3 = 6.67 pies, caterer: 20 * 1/6 = 3.33 pies. Remaining: 20 - (5 + 6.67 + 3.33) = 5 pies.

12. **(A) $7.50**
Explanation: She saved $1.50 per case and bought five cases, so she saved $7.50.

13. **(C) 22.50 minutes**
Explanation: If 8 blocks take 30 minutes, 6 blocks will take (6/8) * 30 = 22.50 minutes.

14. **(B) 500,000**
Explanation: If 600,000 is 1/5 more than predicted, then 600,000 / 6 * 5 = 500,000 was predicted.

15. **(D) 66 2/3 %**
Explanation: She must get 40 out of 60 right, so 40 / 60 = 66 2/3%.

16. **(C) $3,180**
Explanation: 6% of $3,000 = 0.06 * $3,000 = $180. $3,000 + $180 = $3,180.

17. **(B) 14**
Explanation: 3.5 miles / 0.25 miles per lap = 14 laps.

18. **(A) 47**
Explanation: 75 kilometers * 5/8 = 46.875 miles, which is approximately 47 miles per hour.

19. **(C) $574.66**
Explanation: Regular pay = 40 * $12.30 = $492. Overtime pay = 6 * 1.12 * $12.30 = $73.80. Total pay = $574.66.

20. **(A) 6 square feet**
Explanation: Initial space per employee = 42 square feet. New space per employee with 35 employees = 1260 / 35 = 36 square feet. Difference = 42 - 36 = 6 square feet.

21. **(B) $29,000**
Explanation: $450 * 12 * 5 + $2,000 = $29,000.

22. **(D) $114.56**
23. **(C) 33 hours**
24. **(A) two 25-gallon buckets at $550 each**
25. **(B) 200 residents**
26. **(D) 375 miles**
27. **(A) 15 feet**
28. **(A) $16.15**
29. **(C) 16**
30. **(B) $45.12**

Word Knowledge Answers:

1. A.
Apex is a noun that means the top or highest part of something, especially when it forms a point.

2. A.
Assimilate is a verb that means to take in and understand or to cause something to resemble another thing.

3. B.
Brittle is an adjective that means hard but liable to break easily.

4. C.
Datum is a noun that refers to a piece of information.

5. C.
Proficient is an adjective that means competent or skilled in doing (or using) something.

6. C.
Impose is a verb that means to force acceptance of something unwelcome or unfamiliar.

7. A.
Galore is an adjective that means in abundance.

8. D.
Headlong is an adverb and an adjective, and it means with the head foremost or in a reckless rush.

9. D.
Frugal is an adjective that means sparing or economical, particularly regarding money or food.

10. B.
Stimulate is a verb that means to raise levels of physiological or nervous activity within the body or another biological system. It also means to encourage interest or activity in someone or something or to encourage development of increased activity in a state or process. Therefore, the opposite is to arrest, or stop.

11. B.
Illicit is an adjective that means forbidden by law, rules, or customs.

12. D.
Vacate is a verb that means to leave a place that was previously occupied.

13. B.
Reasoned is an adjective that means underpinned by logic or sense.

14. B.
Tacit is an adjective that means understood or implied.

15. D.
Burnish is a verb that means to polish something by rubbing.

16. A.
Commodity is a noun that means a raw material or agricultural product that can be bought or sold; it can also refer to a useful or valuable thing.

17. A.
Contrived is an adjective that means deliberately created instead of arising naturally or spontaneously.

18. C.
Supplicate is a verb that means to ask or beg for something humbly or earnestly.

19. A.
Hypocrisy is a noun that means the practice of claiming standards or beliefs without actually conforming to them. The word most opposite in meaning is honesty.

20. D.
Succulent is an adjective that refers to food being juicy, tender, and delicious.

21. D.
Garrison is a noun that refers to troops stationed in a fortress or town to defend it.

22. A.
Furtherance is a noun that refers to the advancement of an interest or scheme.

23. C.
Domicile is a noun that means a permanent home or a place in which someone lives and has a substantial connection with; it often refers to countries (particularly as it relates to U.S. laws), but not always.

24. D.
Abrogate is a verb that means to repeal or do away with.

25. B.
Compensation is a noun that refers to something awarded to someone as reimbursement for loss, injury, or suffering. It can also refer to the money an employer gives to an employee as a salary or wages.

26. D.
Brusque is an adjective that means abrupt or offhand in speech or manner.

27. B.
Demise is a noun that means someone's or something's death.

28. B.
Commemorate is a verb that means to recall and show respect for someone or something. It also means to celebrate or serve as a memorial to something.

29. C.
Bore is a verb that means to make a hole in something or to make one's way through something.

30. B.
Prevail is a verb that means to be victorious, to be widespread, or to be current.

31. C.
Defray is a verb that means to provide money to pay a cost or expense.

32. A.
Chasm is a noun that refers to a deep fissure in the earth, rock, or another surface; it can also refer to a vast difference between two things.

33. C.
Fundamental is an adjective that means forming a needed base or core; it also refers to a central or main rule or principle on which something is based.

34. D.
Susceptible is an adjective that means likely to be influenced or harmed by something.

35. A.
Emblem is a noun that means a device or symbolic object used as a badge or something that serves as a symbolic representation of a characteristic or concept.

Paragraph Comprehension Answers:

1. C.
Effective appointment management eliminates the waste of your time, as the last sentence of the passage explains.

2. D.
The passage gives the numbers of senators and representatives, so Choice (A) is incorrect. The passage states that each state's population determines the number of representatives a state has, so Choice (B) is incorrect. As the passage states, each state has at least one representative, so Choice (C) is incorrect.

3. A.
Many languages are excluded from the Indo-European language group, so Choice (B) is incorrect. Indians, Iranians, Asians, and Afghans aren't Europeans, so Choice (C) is incorrect. The passage gives no evidence to support Choice (D), which isn't true.

4. D.
Privatized functions operate independently of the government, making Choices (A) and (B) incorrect. The passage states that privatized functions may sell goods and services to other customers and the government, so Choice (C) is also incorrect. Choice (D) is the correct answer because privatized functions perform essential services to government agencies.

5. C.
Choice (A)—"Lead by Example"—is a good philosophy but isn't pertinent to the passage's main point. Choices (B) and (D) are subpoints that support the main point of the passage: how to lead a successful conference, Choice (C).

6. A.
You can assume that causing rain or snow would end a drought, Choice (A). Nothing in the passage has to do with expense, so Choice (B) is incorrect. The passage says nothing about how frequently the process is used, so Choice (C) is incorrect. The passage specifies that dry ice (solid carbon dioxide) is used; regular ice (solid water) is a different substance, so Choice (D) is wrong.

7. C.
Choices (A) and (B) may be true in certain situations, but they're not the point of this particular paragraph. The passage says nothing about working to improve writing skills being a waste of time, so Choice (D) is incorrect. Instead, the paragraph's main point is that writing may not be the most efficient communication method, depending on the situation.

8. B.
Products with transistors weren't widely sold before 1954 because of the expense and difficulty of production, not because markets didn't exist, so Choice (A) is incorrect. Choice (c) has the right date, but the passage doesn't say who invented the transistor, so it's wrong as well. Choice (D) is wrong because the passage states that the problem of transistors being expensive to produce was solved in 1954. The last sentence notes that the first transistor radio went on the market after cost and production problems were overcome, so Choice (B) is the right answer.

9. B.
The words *London* and *Englishman* make it clear that the author is speaking of his travels in England, which is part of Great Britain.

10. A.
The author makes no reference to parents in the passage, so Choice (B) is incorrect. The author doesn't imply that all these interests require equal time, so Choice (C) is incorrect. The passage is about children under 18; you can't draw a conclusion about what the author thinks people over 18 should do, so Choice (D) is incorrect.

11. C.
The passage doesn't say anything about mock job interviews being frightening, so Choice (A) is wrong. The passage says that mock job interviews should be available to all students, so Choice (B) is wrong. The passage says that schools, not students, should organize mock interviews, so Choice (D) is incorrect.

12. A.
Choices (B), (C), and (D) are the opposite of what the paragraph states and implies.

13. B.
Nothing in the paragraph supports Choice (A), which is incorrect. When an amendment is passed, it becomes part of the Constitution, so Choice (C) is incorrect. The passage doesn't support Choice (D) because the passage doesn't mention anything related to lottery tickets. The passage defines *due process* as "the guarantee of fairness in the administration of justice," so Choice (B) is correct.

14. C.
Because the 14th Amendment guarantees due process in states' laws, the 5th Amendment must guarantee due process only in federal law, which makes Choice (C) right. Nothing in the passage implies that the 5th Amendment is about taxes, so Choice (A) is wrong. Because the passage states that the 14th Amendment had to be enacted to require states to abide by due process, Choice (B) is incorrect. Choice (D) is neither stated nor implied in the passage.

15. A.
Because the Supreme Court struck down many state laws after the 14th Amendment was enacted, it's probably true that these laws would still exist if there had been no 14th Amendment. The passage doesn't support Choices (B), (C), or (D).

Mathematics Knowledge Answers:

1. **B. 150 miles.**
Explanation: The distance a vehicle travels is equal to its speed times the time it travels. So, 60 miles/hour * 2.5 hours = 150 miles.

2. **C. 6.**
Explanation: To solve for x, subtract 4 from both sides to get 3x = 18, then divide both sides by 3 to get x = 6.

3. **B. 6.**
Explanation: Add 3 to both sides to get 5x = 30, then divide both sides by 5 to get x = 6.

4. **B. $31,250.**
Explanation: If $25,000 is 80% of the original price, then the original price is $25,000 / 0.80 = $31,250.

5. **D. 12 inches.**
Explanation: If the length of the rectangle is twice the width, then the rectangle has dimensions of 2x and x. The perimeter of the rectangle is 2*(2x + x) = 36, so 6x = 36 and x = 6. Therefore, the length of the rectangle is 2*6 = 12 inches.

6. **B. 720 degrees.**
Explanation: The sum of the interior angles of a polygon with n sides is 180(n - 2). For a hexagon, n = 6, so the sum of the interior angles is 180(6 - 2) = 720 degrees.

7. **B. 0.8x.**
Explanation: If a number is decreased by 20% of itself, it is left with 80% of its original value, which is 0.8x.

8. **D. 125 cubic cm.**
Explanation: The volume of a cube is given by side^3. So, 5^3 = 125 cubic cm.

9. **C. 40.**
Explanation: The mean of a set of numbers is the sum of the numbers divided by the number of numbers. If the mean of four numbers is 10, their sum is 4*10 = 40.

10. **A. 2.**
Explanation: The slope-intercept form of a line is y = mx + b, where m is the slope. To get this form from 3y = 6x + 12, divide both sides by 3 to get y = 2x + 4. The slope of the line is 2.

11. **B. -2.**
Explanation: Subtract 2x from both sides to get -3 = 2x + 1. Then, subtract 1 from both sides to get -4 = 2x. Finally, divide both sides by 2 to get -2 = x.

Answers and Explanations | 183

12. **A. 81.**
Explanation: 3^4 is 3*3*3*3, which equals 81.

13. **A. 18 hours.**
Explanation: When both the inlet and outlet are open, the net rate of filling the tank is the difference between the filling and emptying rates, which is 1/6 - 1/9 = 1/18 tanks per hour. Therefore, it will take 18 hours to fill the tank.

14. **C. 16.**
Explanation: The square root of 256 is 16.

15. **D. 6.**
Explanation: Subtract 2y from both sides to get 2y - 7 = 5. Then, add 7 to both sides to get 2y = 12. Finally, divide both sides by 2 to get y = 6.

16. **B. 20 mpg.**
Explanation: The car's fuel efficiency is calculated as the distance traveled divided by the fuel consumed. So, 400 miles / 20 gallons = 20 miles per gallon (mpg).

17. **B. 10 units.**
Explanation: The area of a rectangle is length times width. If the area is 100 square units and the length is 10 units, the width is 100 / 10 = 10 units.

18. **B. 10.**
Explanation: 5% of 200 is 0.05 * 200 = 10.

19. **B. $25.**
Explanation: If the shirt is sold for $20 after a 20% discount, then $20 is 80% of the original price. So, the original price is $20 / 0.80 = $25.

20. **D. 120.**
Explanation: The least common multiple (LCM) of 6, 8, and 10 is 120.

21. **B. 8.**
Explanation: If the ratio of cats to dogs is 3:2, then for every 5 animals, there are 3 cats and 2 dogs. Since there are 20 animals, this ratio will occur 4 times. Therefore, there are 2 * 4 = 8 dogs.

22. **D. 40.**
Explanation: The first 5 terms of the sequence are 2, 5, 8, 11, and 14. Their sum is 2 + 5 + 8 + 11 + 14 = 40.

23. **B. 2.**
Explanation: Subtract 3 from both sides of the equation to get 2x = 4. Then divide both sides by 2 to get x = 2.

24. **B. 1.15x.**
Explanation: If a number is increased by 15% of itself, the new number is 1.15 times the original number, or 1.15x.

25. **B. $400.**
Explanation: If $300 is 75% of the original price, the original price is $300 / 0.75 = $400.

26. **A. 540 degrees.**
Explanation: The sum of the interior angles of a polygon with n sides is 180(n - 2). For a pentagon, n = 5, so the sum of the interior angles is 180(5 - 2) = 540 degrees.

27. **B. 13.**
Explanation: Substituting a = 2 and b = 3 into the expression gives 2^2 + 3^2 = 4 + 9 = 13.

28. **B. 1/3.**
Explanation: There are 2 outcomes (5 and 6) greater than 4 out of 6 total outcomes, so the probability is 2/6 = 1/3.

29. **A. -8.**
Explanation: Subtract 2y from both sides to get y + 5 = -3. Then, subtract 5 from both sides to get y = -8.

30. **C. 3.**

Electronics Information Answers

1. A.
Ohm's law states that Voltage(V OR E) = Current (I)x Resistance(R). All other answers are incorrect expressions of this law.

2. A.
You read a resistor's color bands from left to right. The first band denotes the first digit, the second band denotes the second digit, and the third band denotes the subsequent number of zeros. So in this example, brown is one, black is zero, and red means two additional zeros.

3. B.
Heaters, pumps, stairs, diving boards, railings, and rebar, among other things, must be bonded together by a minimum #8 wire for safety purposes.

4. D.
Voltage is commonly used as a short name for electrical potential difference, measured in volts.

5. A.
SIM stands for *Subscriber Identity Module*. The card contains information such as your phone number, your billing information, and your address book. The card makes it easier to switch from one cell phone to another.

6. A.
A resistor is so named because it resists (or inhibits) the current flow.

7. C.
A diode has two terminals, the anode and the cathode, which is why it's called a *diode*. It restricts current flow to only one direction.

8. C.
Radar can operate as high as 100,000 MHz (megahertz).

9. B.
Current is the flow of charged particles. The difference between alternating current (AC) and direct current (DC) is that the electrons in an AC circuit regularly reverse their direction. In a DC circuit, electrons always flow in the same direction.

10. C.
Changing AC to DC is a process called *rectification*.

11. C.
A British thermal unit (BTU) is a measure of heat energy.

12. D.
Current is the presence of electron flow.

13. B.
Transducers, which transform or convert energy, can be switches, strain gauges, temperature sensors, or inductive switches. A transformer is an inductor that increases or decreases voltage.

14. D.
Active elements are electronic devices that can create energy (such as voltage and current supplies). *Passive elements* are electronic devices that cannot create energy.

15. A.
Power= Current x Voltage or, written another way, current= Power/Voltage. Plug in the numbers and do the math: 60 watts/120volts= 0.5 amperes.

16. C.
The larger the number, the smaller the diameter of the wire.

17. C.
Because fuses are designed to prevent current overload at a specific level, a fuse with a high rating may allow a higher current to flow through a circuit not designed to work at that higher current, possibly causing damage to the circuit.

18. A.
Neutral wire is always whitish or naturally colored.

19. D.
Electrical power is measured in watts, so you use a wattmeter. An ammeter measures amps (current). An ohmmeter measures ohms (resistance). A voltmeter measures volts (voltage).

20. B.
The bulb will burn dimmer because its full potential isn't used; it'll last longer for the same reason.

Auto & Shop Information Answers:

1. D.
Overheating can melt engine parts, enlarge pistons, and burn engine bearings.

2. B.
Alternators convert mechanical energy (rotary motion) into electrical energy (output current).

3. B.
Electronic ignition systems use lower input voltages to get higher output voltages (for spark).

4. A.
Piston rings are seals that keep the exploding gases in the combustion chamber.

5. A.
The crankshaft is connected to the flywheel, which causes it to rotate, operating the pistons.

6. A.
A differential lets wheels turn at different rates. So when one tire is spinning, the other can have traction.

7. D.
The ignition system starts a car. If there doesn't seem to be any electricity, check the battery.

8. C.
You use a coping saw to make intricate cuts in wood.

9. D.
Snowmobiles, chainsaws, lawnmowers, and some motorcycles need only two-stroke engines.

10. B.
You often use a belt sander is often used to finish wood because doing so is faster than hand sanding for large areas.

11. C.
A limited-slip differential transfers the most driving force to the wheel with the greatest Traction.

12. A.
Big block engines generally have greater than 5.9-liter displacement.

13. C.
A trowel would be a good tool for spreading and/or shaping mortar.

14. B.
You use a plumb bob to check vertical reference using a pointed weight on a line.

15. B.
Rebar is an iron bar that is embedded in cement to reinforce it.

16. C.
Annular ring, clout, and spring head are all types of nails.

17. B.
A ripsaw cuts wood with the grain, called *ripping*, because it's easier than cutting against the grain.

18. B.
A cam belt, also known as a *timing belt*, connects the crankshaft to the camshaft.

19. C.
You use a level to check for horizontal trueness.

20. C.
A bucking bar is used to form rivet bucktails.

21. A.
Washers that have teeth all around the circumference are called shake-proof washers.

22. D.
The tool shown is an outside caliper measuring wire thickness and other objects.

23. A.
The object pictured is a nut.

24. D.
The tool is a plane; you use it to dress (prepare, smooth, and shave) wood.

25. A.
Commonly called a cold chisel, this tool cuts through metal screws, bolts, nails, and other metals.

Mechanical Comprehension Answers:

1. A.
An induction clutch is a magnetic clutch. When a conductor (wire) is wrapped around a core and electricity passes through the wire, it creates a magnetic field. The same wire also acts as an inductor, which produces inductance during AC current flow. It's similar to resistance in a resistor in that it "resists" current flow, but the value of inductance is based on the value of the inductor (written as *L*) and the frequency of the AC current. Therefore, an induction clutch uses magnetism to operate.

2. B.
You can calculate mechanical advantage as Length of Effort Arm Length of Resistance Arm. Plug in the numbers: $MA=8/2=4$.

3. C.
The box with the largest area on the bottom will have the shortest sides. If LengthX Width X Height= Volume and all the boxes have equal volume, then the sides must be shortest on the box with the largest area on the bottom. Calculate the area of each box Bottom:

No. 1 = 20 square inches
No. 2= 35 square inches
No. 3=48 square inches
No. 4 =27 square inches
No. 3 which has the largest area, will have the shortest sides.

4. C.
Anvil B's landing on the seesaw will propel Anvil A into the air

Answers and Explanations | 187

5. A.
Pressure equals force divided by area in square inches ($P = F \times A$). You can also state this formula as $F = A \times P$. Substitute the known quantities: $F = 15 \times 24 = 360$ pounds.

6. C.
Recoil occurs when an object producing a force is kicked back.

7. A.
To determine the answer, multiply the number of teeth Gear 1 has (D) and the number of revolutions it makes (R). Divide that number by the number of teeth Gear 2 has (d) to determine the number of revolutions Gear 2 makes (r). Because the gears are proportional, this formula shows you the ratio of teeth to revolutions.

$$r = \frac{DR}{d}$$
$$= \frac{25 \times 10}{15}$$
$$= \frac{250}{15} = \frac{50}{3} = 16\frac{2}{3}$$

That means Gear 2 (the smaller gear) makes $16\frac{2}{3}$ revolutions for every 10 revolutions that Gear 1 (the larger gear) makes.

8) B.
You can determine the pressure of all that water by multiplying the volume of the aquarium by the weight of the water—volume *lwh*. The bottom of the tank is 18 feet long by 12 feet wide by 10 feet high for a total volume of 2,160 cubic feet: $18 \times 12 \times 10 = 2160 \text{ft}^3$. A cubic foot of water weighs approximately 62.5 pounds, so multiply the volume of water by 62.5: $2160 \times 62.5 = 135,000$. That gives an approximate pressure on the bottom of the tank of about 135,000 pounds over the entire surface area. The surface area of the bottom of the tank is length width. Convert feet to inches and then find the area: $A(18\text{ft.} \times 12\text{in/ft}) \times (12\text{ft}/12\text{ft/in.}) = 216\text{in} \times 144\text{in}$. $= 31104 \text{ in}^2$ Dividing the pressure of 135,000 by the number of square inches of surface area gives an approximate psi of 4.

9. D.
Machine springs are usually made of steel, although sometimes they're made of brass or other metal alloys.

10. B.
Clutches connect and disconnect parts, so they're a type of coupling.

11. D.
When the high point of the cam connects with the lever arm, the lever arm will touch the contact point. Two high points on the cam mean the lever arm will touch the contact point twice with each revolution of the cam.

12. C.
A single moveable block-and-fall is called a runner. A runner provides a way to get mechanical advantage by threading a rope through a pulley attached to the load, anchoring one end to a stationary point, and you pulling on the other end of the rope, hoisting the load.

13. C.
If the fulcrum is moved closer to the anvil, the length of the effort arm of the lever will be increased, making the anvil easier to raise, but the height to which the anvil can be raised will be reduced.

14. B.
In this block-and-tackle arrangement, there are three segments coming off the load, which gives this arrangement a mechanical advantage of 3.

15. C.
The larger cog (Cog A) covers a greater linear distance in a given period of time, so Cog A reaches the top first.

16. A.
The key will feel coldest because metal is a better conductor than the other materials.

17. D.
All but Valve 4 should be open. Opening Valves 1 and 2 allows water to enter the tank. Opening Valves 3 and 5 prevents water from filling the tank entirely. Opening Valve 4 allows water to leave the tank.

18. A.
Gears with their teeth together in mesh turn in opposite directions. Gear A turns Gear B in the opposite direction (right), and Gear B turns Gear C in the opposite direction (left).

19. A.
Gears with their teeth together in mesh turn in opposite directions. Gear 1 turns clock wise. Gear 2, in mesh with Gear 1, turns counterclockwise. Gear 3, in mesh with Gear 2, turns clockwise. Gear 4, in mesh with Gear 3, turns counterclockwise. Gear 5, in mesh with Gear 2, turns clockwise.

20. C.
The gauge shows a reading of 21.

21. D.
The formula for determining power is Power Work Time.

22. C.
Silver is the best conductor, so it will become hotter faster than the other objects because heat transfers faster into materials with greater conductivity than those with lower Conductivity.

23. A.
Stationary pulleys (often called whips) give no mechanical advantage, so effort equals the weight of the crate or 50 pounds.

24. C.
At the height of the arc, the ball has no upward momentum, so it goes the slowest at that point (it has zero upward movement at its highest point).

25. A.
The brace on Angle A covers more area of the angle, so it's more solidly braced.

Assembling Objects Answers:

1. C.
Connect Point A on the star and Point A on the line. Next, rotate the star-shaped figure about 45 degrees and connect Point B on the line to Point B on the shape, rotating the shape 90 degrees.

2. A.
Count the number of shapes in the original image (there are seven). Only one answer choice includes the seven shapes in the original figure. In this case, Choices (B), (C), and (D) just don't have enough shapes—and that makes it easy to get this question correct.

3. D.
Point B in the original image is on the end of the crescent, which means you can automatically rule out Choice (C), in which the point is in the center of the figure. Choices (A) and (B) are also incorrect because, in the original figure, the line goes directly through the point of the triangle. Therefore, the only answer that reflects that intersection is Choice (D).

4. D.
There are three shapes in the original puzzle-like diagram, so you can automatically rule out Choices (A) and (B)—they each have four. However, if you rotate the shapes depicted in the question, you'll see that two sections are larger than the remaining one. Therefore, the only reasonable answer is Choice (D) because it shows an oval inside a triangle with more space in two corners than in the remaining corner.

5. A.
Rotate the pentagon to the left so Point A is on the top. Flip the other shape almost 180 degrees so Point B is on the bottom. Connect the two shapes with the line's corresponding points, and you'll have Choice (A) on your hands. If you're unsure, you can always rule out answers that you know aren't correct. In this case, you can tell Choice (B) is wrong because Point B is in the wrong place; it doesn't match the original diagram. Likewise, choice (C) is incorrect because neither point is in the correct places, and Choice (D) is wrong for the same reasons.

6. A.
Reconfigure the shapes in the original figure so that the two largest are side-by-side and symmetrical, with the uppermost points touching. The remaining two shapes fit right inside, like a puzzle. (On this question, you can immediately rule out *all* the other choices because the shapes don't match the original figure; Choices (C) and (D) even depict too many shapes to be contenders.)

7. C.
On Assembling Objects questions, pay attention to the original figures. In this question, the line in Shape B starts in the center of the triangle and travels through one of the triangle's points. The only answer choice that accurately depicts Figure B is Choice (C), so you can ignore all the other choices.

8. D.
Start by counting the number of shapes in the original diagram—six. Remember that counting shapes doesn't always work to rule out incorrect answers, but you can tell (in this problem, at least) that Choice (B) is incorrect because it depicts only four shapes. If you can't picture the correct answer, go through the other choices to eliminate the wrong ones. Choice (A) can't be correct because it depicts two triangle-like shapes that aren't in the original diagram. Choice (C) can't be correct because the shapes it depicts aren't even similar to those depicted in the question, which leaves Choice (D) as the only possible correct answer.

9. B.
Choice (B) is correct because Point A on the circle and Point B on the triangle match the original diagram, as does the line length. However, choice (A) can't be right because Figure A is flipped, the line connecting the two figures is too short, and Choice (C) shows Point B in the wrong place on the triangle. Choice (D) is wrong, too, because Point A is on the wrong circle.

10. B.
The shapes in the original diagram are all very similar (and this is one case where counting the number of shapes won't work). However, the shapes in Choices (A), (C), and (D) don't match those in the original diagram, leaving Choice (B). If you rearrange the three trapezoids in your mind by connecting the points on the longest side of each, you'll see Choice (B) take shape.

11. A.
Point A is on the long, flat side of the first shape (the one with four points), whereas Point B is just about in the center of the second shape (the one with five points). When you look at the answer choices, you'll see that Choices (B) and (D) depict the points in the wrong places, so those choices are automatically wrong. Choice (C) looks like it has the points in the correct places, but not so fast—they're attached to the wrong shapes.

12. D.
Compare the shapes in the question with the shapes in the answer choices. You can see that Choice (A) is incorrect because the shapes don't match; so is Choice (C). The shapes in the question include three very similar figures, but two are large, and one is small. Choice (B) can't be correct because two of the shapes are small, and one is large, creating an oval tucked into the bottom of a triangle. Choice (D) is the right answer because it takes two larger shapes and one smaller shape to create an upright oval inside a triangle.

13. D.
Both of the figures in the original diagram are the same; they're just mirror images of each other. Choice (D) is correct because it's the only figure that shows them as mirror images of each other with the points in the correct places.

14. C.
You can immediately put Choice (A) in the recycle bin—it depicts four triangles. The same applies to Choice (D), except it depicts a square and three rectangles. Choice (B) shows five triangles, which can't be correct, either (the original diagram shows four shapes and only three are triangles). Choice (C) is the only one that can be correct.

15. B.
The first thing you should do on a problem like this one is check out where the points are located on the shapes in the original diagram. On this question, you can rule out Choices (A), (C), and (D) because the points are in the wrong places.

16. A.
One of the figures in the original diagram is a triangle with a line running parallel to one side. Choices (B) and (C) don't show such shapes, so those are out the window. Choice (D) has too many triangles (there are six, but the original figure depicts only four), leaving Choice (A) as the correct answer.

17. B.
Choices (C) and (D) are incorrect because they each feature shapes that aren't depicted in the original diagram. Choice (A) is also incorrect; although it's a mirror image of the correct answer, Choice (B) requires you to flip the shapes rather than rotate them.

18. B.
Look closely at where the points are in the original diagram—particularly Point B, which is on the corner of the square. You know that Choice (D) isn't the answer because it shows Point B inside the square. Choices (A) and (C) are wrong, too, because they require you to flip the *F* shape, whereas Choice (B) doesn't even require you to rotate it.

19. C.
Choice (C) is the only choice in which each shape corresponds with a shape in the original diagram.

20. A.
The only answer choice that doesn't require you to invert a shape and doesn't feature a point in the wrong place on one of the shapes is Choice (A). Choice (B) requires you to flip the shape with Point A on it, and so does Choice (D). Choice (C) has the point on the wrong place inside the hexagon.

21. B.
Choice (A) depicts four shapes (the smallest of which isn't in the original diagram), so you can rule that one out immediately. Choices (C) and (D) don't use any five-sided shapes, but there's one in the original diagram. Choice (B) is the only figure that uses replicas of the three shapes in the original illustration.

22. D.
Point B is marked in the original illustration, but in Choices (A), (B), and (C), it's in all the wrong places; that leaves you with Choice (D) as the correct answer.

23. A.
Look closely at Choice (C)—it has a part missing, so you know it can't be correct. Choice (A) is the only one that makes use of all the shapes depicted in the original diagram;

Choices (B) and (D) use shapes that aren't in the question.

24. C.
Point A is located in the lower-right corner of the original figure in the question, so you can ignore Choices (A), (B), and (D) because they don't match.

25. C.
Check each shape to see which answer choices it appears in before you try to put together this puzzle. The only answer choice that features the right shapes in the right proportions (those that match the original illustration) is Choice (C). All the other answers have odd shapes that don't match.

CHAPTER 15

Practice Exam 2

General Science:

1. What is the change in body form that some insects undergo from birth to maturity?
(A) transformation
(B) metamorphosis
(C) trinity
(D) transmutation

2. An earthquake that measures 4 on the Richter scale would be how many times stronger than an earthquake that measured 2?
(A) 2 times stronger
(B) 4 times stronger
(C) 10 times stronger
(D) 100 times stronger

3. Muscles attach to bone with:
(A) non-connective tissue.
(B) ligaments.
(C) tendons.
(D) rubber bands.

4. The male part of a flower is called:
(A) the stamen.
(B) the pistil.
(C) the throttle.
(D) stubborn.

5. Blood leaving the lungs is:
(A) hydrogenated.
(B) coagulated.
(C) watery.
(D) oxygenated.

6. Of these, which river is the longest?
(A) Mississippi
(B) Nile
(C) Colorado
(D) Congo

7. The branch of science that studies matter and energy is called:
(A) chemistry.
(B) physics.
(C) oceanography.
(D) trigonometry.

8. Which type of cloud's name comes from the Latin word meaning "rain"?
(A) nimbus
(B) cirrus
(C) strato
(D) alto

9. Deoxyribonucleic acid is better known as:
(A) antacid.
(B) carbohydrates.
(C) triglyceride.
(D) DNA.

10. The instrument used to measure wind speed is:
(A) barometer.
(B) anemometer.
(C) altimeter.
(D) fanometer.

11. Electric charges can be:
(A) positive or negative.
(B) positive or neutral.
(C) negative or neutral.
(D) neutral only.

12. Which planet in the solar system has the most moons?
(A) Neptune
(B) Saturn
(C) Jupiter
(D) Uranus

13. The law of universal gravitation was discovered by:
(A) Albert Einstein.
(B) Isaac Newton.
(C) Alexander Graham Bell.
(D) Rod Powers.

14. Which U.S. space program is responsible for putting 12 men on the moon?
(A) Gemini
(B) Titan
(C) Voyager
(D) Apollo

15. Animals that eat both plants and animals are Called:
(A) herbivores.
(B) carnivores.
(C) omnivores.
(D) ambidextrous.

16. Unlike most other fish, sharks have no:
(A) gills.
(B) bones.
(C) liver.
(D) heart.

17. What human organ is responsible for the detoxification of blood?
(A) liver
(B) kidneys
(C) intestines
(D) stomach

18. Kinetic energy is the energy that:
(A) is produced by sound waves.
(B) an object potentially has.
(C) is possessed by a moving object.
(D) results from the attraction of two Magnets.

19. The terrestrial planets consist of:
(A) Jupiter, Saturn, Uranus, and Neptune.
(B) Pluto and Neptune.
(C) Mercury, Venus, Earth, and Mars.
(D) any planet.

20. An example of a mineral is:
(A) calcium.
(B) vitamin C.
(C) granite.
(D) almonds.

21. Which animal has the heaviest brain?
(A) human
(B) elephant
(C) rhinoceros
(D) sperm whale

22. The sun is what type of star?
(A) O type
(B) G type
(C) F type
(D) M type

23. Molecules are created when:
(A) matter is created.
(B) matter is destroyed.
(C) atoms combine together.
(D) atoms are separated.

24. An example of an embryonic plant would be a
(A) tree.
(B) rose.
(C) seed.
(D) cabbage.

25. The vernal equinox is
(A) the first day of winter.
(B) near the equator.
(C) the first day of spring.
(D) a lunar eclipse.

Arithmetic Reasoning:

If the sum of two numbers is 100 and their difference is 50, what is the smaller number?
A. 25
B. 50
C. 75
D. 100

120 miles on 5 gallons of gas. How many miles can it travel on 9 gallons?
A. 180 miles
B. 216 miles
C. 240 miles
D. 270 miles

A train travels at 60 mph for 2 hours and 30 minutes, then at 80 mph for another 1 hour and 30 minutes. What was its average speed for the entire trip?
A. 64 mph
B. 66 mph
C. 68 mph
D. 70 mph

The area of a rectangle is 84 square inches. If the length of the rectangle is twice its width, what is the length of the rectangle?
A. 6 inches
B. 8 inches
C. 10 inches
D. 12 inches

If 30% of a number is 150, what is 80% of that number?
A. 300
B. 400
C. 500
D. 600

If a shirt is on sale for 25% off and the sale price is $45, what was the original price?
A. $50
B. $55
C. $60
D. $65

A manufacturer sells widgets for $5 each. If it costs $3 to make each widget, how many widgets does the manufacturer need to sell to make a profit of $1000?
A. 200
B. 500
C. 600
D. 800

If the sales tax rate is 7.5%, how much tax will you pay on an item that costs $200?
A. $10
B. $15
C. $20
D. $25

A clock gains 5 minutes every hour. If it was set correctly at noon, what time will it show at 6 PM?
A. 6:15 PM
B. 6:30 PM
C. 6:45 PM
D. 7:00 PM

If a worker earns $20 an hour for the first 40 hours in a week and time and a half for any hours over 40, how much will he earn in a week where he works 50 hours?
A. $800
B. $900
C. $1000
D. $1100

A factory produces 1000 widgets per day. If the defect rate is 0.5%, how many defective widgets are produced in a week?
A. 35
B. 50
C. 75
D. 100

If a salesman earns a 10% commission on all sales, how much will he earn on $15,000 of sales?
A. $1,500
B. $1,750
C. $2,000
D. $2,250

If a car depreciates by 15% each year, what will be the value of a car initially worth $20,000 after 3 years?
A. $12,000
B. $14,000
C. $16,000
D. $18,000

A football field is 100 yards long and 50 yards wide. What is the area of the football field in square feet (1 yard = 3 feet)?
A. 15,000 sq ft
B. 30,000 sq ft
C. 45,000 sq ft
D. 60,000 sq ft

If a box contains 24 cans of soda and each can hold 12 ounces of soda, how many gallons of soda are in the box (1 gallon = 128 ounces)?
A. 2.25 gallons
B. 3.25 gallons
C. 4.5 gallons
D. 5 gallons

If a recipe requires 3 cups of flour to make 2 dozen cookies, how many cups of flour would be needed to make 5 dozen cookies?
A. 6.5 cups
B. 7.5 cups
C. 8.5 cups
D. 9.5 cups

If the ratio of boys to girls in a class is 3:2 and there are 20 students in the class, how many boys are there in the class?
A. 8
B. 10
C. 12
D. 14

If a man's salary is $40,000 and he gets a 5% raise, what will his new salary be?
A. $42,000
B. $42,500
C. $43,000
D. $43,500

A car travels at an average speed of 45 miles per hour. How many minutes will it take to travel 15 miles?
A. 10 minutes
B. 20 minutes
C. 30 minutes
D. 40 minutes

How many square feet of carpeting is needed to carpet a 10-foot-x-10-foot room?
A. 500
B. 100
C. 150
D. 200

Word Knowledge:

1. Abeyance most nearly means:
(A) trustworthiness.
(B) passion.
(C) suspension.
(D) business.

2. It was a sturdy table.
(A) well-built
(B) ugly
(C) thick
(D) small

3. Bullock most nearly means:
(A) ox.
(B) inattentive.
(C) lazy.
(D) panther.

4. Brevity is the soul of wit.
(A) beauty
(B) intelligence
(C) clarity
(D) humor

5. Paradigm most nearly means:
(A) twenty cents.
(B) model.
(C) heaven.
(D) basis.

6. My manger facilitated my promotion.
(A) hindered
(B) helped
(C) disliked
(D) ignored

7. Quiescence most nearly means:
(A) kill.
(B) preserve.
(C) small.
(D) quiet.

8. The spectator enjoyed the game.
(A) competitor
(B) observer
(C) referee
(D) organizer

9. Joy reclined against the far wall.
(A) sat
(B) leaned
(C) jumped
(D) paraded

10. The teacher cited some examples.
(A) memorized
(B) finished
(C) quoted
(D) examined

11. Surround most nearly means:
(A) line.
(B) benefit.
(C) encircle.
(D) speaker.

12. Illustrious most nearly means
(A) illustrated.
(B) famous.
(C) foolish.
(D) intelligent

13. Inhabitant most nearly means:
(A) invalid.
(B) nun.
(C) seeker.
(D) dweller.

14. Tim had a penchant for engaging in subterfuge.
(A) religion
(B) intrigue
(C) gambling
(D) danger

15. Megan found the adults' costumes to be ghastly.
(A) hideous
(B) cute
(C) large
(D) comfortable

16. Rigid most nearly means:
(A) strong.
(B) weak.
(C) pliable.
(D) inflexible.

17. Billy yearned to join the fraternal organization.
(A) brotherly
(B) large
(C) fun
(D) special

18. Deplore most nearly means:
(A) accept.
(B) insult.
(C) disapprove.
(D) salute.

19. Meager most nearly means:
(A) space.
(B) sparse.
(C) brief.
(D) thirsty.

20. Weal most nearly means:
(A) happiness.
(B) blow.
(C) scream.
(D) tire.

21. To be guileless, I think your hair looks ugly.
(A) helpful
(B) kind
(C) frank
(D) serious

22. The customs agent confiscated the goods.
(A) bought
(B) noticed
(C) seized
(D) stole

23. Dubious most nearly means:
(A) long.
(B) beautiful.
(C) articulate.
(D) doubtful.

24. Illusion most nearly means:
(A) mirage.
(B) distant.
(C) sight.
(D) perspective.

25. Becky developed a sudden craving for ice cream.
(A) disgust
(B) passion
(C) hatred
(D) desire

26. Enmity most nearly means:
(A) enemy.
(B) hostility.
(C) anger.
(D) childish.

27. Arbor most nearly means:
(A) native.
(B) tree.
(C) travel.
(D) delirious.

28. They terminated the contract.
(A) bought
(B) extended
(C) sold
(D) ended

29. Tim always considered Chuck to be a big buffoon.
(A) clown
(B) help
(C) liar
(D) pain

30. Null most nearly means:
(A) zero.
(B) dull.
(C) unskilled.
(D) rapid.

31. Tom had to prove to the judge that he was not indigent.
(A) guilty
(B) rich
(C) poor
(D) ugly

32. Impertinent most nearly means:
(A) fun.
(B) boring.
(C) rude.
(D) impatient.

33. Lustrous most nearly means:
(A) expensive.
(B) lazy.
(C) cold.
(D) polished.

34. Pardon most nearly means:
(A) courtesy.
(B) excuse.
(C) believe.
(D) respect.

35. Veracious most nearly means:
(A) fast.
(B) slow.
(C) equal.
(D) truthful.

Paragraph Comprehension:

Questions 1 and 2 are based on the following passage.

There is not a single town of any size within a distance of forty miles, yet already the rural population of this county is quite large. The whole country, within a wide circuit north, south, east, and west, partakes of the same general character; mountain ridges, half-tilled, half-wood, screening cultivated valleys, sprinkled with farms and hamlets, among which some pretty stream generally winds its way. The waters in our immediate neighborhood flow southward, though only a few miles to the north of our village, the brooks are found running in an opposite course, this valley lying just within the borders of the dividing ridge. The river itself, though farther south it becomes one of the great streams of the country, cannot boast of much breadth so near its source, and running quietly among the meadows, half screened by the groves and thickets, scarcely shows in the general view.

1. According to this passage,

(A) the author lives in a large city.
(B) the author lives in the country.
(C) the author lives on the seashore.
(D) the author lives on Mars.

2. According to this passage, the brooks are running in which direction within the author's neighborhood?

(A) north
(B) south
(C) east
(D) west

The Panama Canal is a ship canal that cuts through the Isthmus of Panama, connecting the Atlantic and Pacific oceans. Although several foreign companies tried to build the canal throughout the 19th century, none were successful. After the U.S. helped Panama revolt against Colombia, the U.S. was given rights to the land the canal occupied. The U.S. government finished the canal in 1914.

3. According to this passage,

(A) Panama and Colombia fought a war over the Panama Canal.
(B) the U.S. was given rights to the canal land.
(C) foreign companies built the canal before the U.S. stepped in.
(D) Panama built the canal in 1914.

Extreme care must be exercised to ensure proper handling and cleaning of soiled U.S. flags. A torn flag may be professionally mended, but a badly torn or tattered flag should be destroyed. When the flag is in such a condition that it's no longer a fitting emblem for display, destroy it in a dignified manner, preferably by burning it.

4. According to this passage, torn flags should be:

(A) mended.
(B) burned.
(C) destroyed.
(D) all of the above.

Medieval guilds were similar to modern-day labor unions. These groups of merchants or craftspeople set economic activity rules to protect themselves. Some guilds held considerable economic power, but even small guilds protected members. Guilds also served a social purpose.

5. According to this passage, guilds:

(A) had only one purpose.
(B) had little in common with modern labor unions.
(C) exploited workers.
(D) held considerable economic power.

After a series of well-publicized failures by various inventors, Orville and Wilbur Wright succeeded in flying and controlling a heavier-than-air craft on December 17, 1903; the War Department, stung by its investment in a failed effort by Samuel Langley and compounded by the Wrights' secretiveness, initially rejected the brothers' overtures toward the government to buy the aircraft. Prevailing sentiments held that the immediate future still belonged to the balloon. In August 1908, the two brothers delivered the first Army aircraft to the U.S. Government. That the U.S. government managed to purchase an airplane was a minor miracle. The government refused to accept that man had flown in a heavier-than-air machine for more than four years after the Wright brothers' successful flight at Kitty Hawk, North Carolina.

6. Which of the following statements is not supported by the above passage?

(A) The U.S. Government felt that balloons were more practical than airplanes.
(B) The Wright brothers' own secretiveness contributed to their problems in getting the government interested in their aircraft.
(C) The historic flight took place on the East Coast.
(D) It took more than six years for the Wright brothers to interest the U.S. Government in their airplane.

If anyone should be inclined to overrate the state of our present knowledge of mental life, all that would be needed to force him to assume a modest attitude would be to remind him of the function of memory. No psychologic theory has yet been able to account for the connection between the fundamental phenomena of remembering and forgetting; indeed, the complete analysis of what one can observe has scarcely been grasped. Today forgetting has perhaps grown more puzzling than remembering, especially since we have learned from the study of dreams and pathologic states that even what for a long time we believed forgotten may suddenly return to consciousness.

7. The primary subject of this paragraph is:

(A) bowling.
(B) puzzles.
(C) memory.
(D) government service.

Troy weight is based on a pound of 12 ounces and an ounce of 480 grains. Common, or avoirdupois, weight is based on a pound having 16 ounces and an ounce having 437.5 grains. A common pound has 7,000 grains, while a troy pound has 5,760.

8. According to this passage,

(A) in common weight, an ounce is less than 438 grains.
(B) a troy pound and a common pound are the same weight.
(C) common weight and avoirdupois weight are different measures.
(D) a troy ounce equals 437.5 grains.

Good leaders get involved in their subordinates' careers. People merely obey arbitrary commands and orders but respond quickly and usually give extra effort to leaders who genuinely care for them. An often neglected leadership principle in today's environment of technology and specialization is knowing the workers and showing sincere interest in their problems, career development, and welfare. Leadership is reflected in the degree of efficiency, productivity, morale, and motivation demonstrated by subordinates. Leadership involvement is the key ingredient to maximizing worker performance.

9. A fundamental leadership principle that's often ignored is:

(A) leading by example.
(B) showing sincere interest in the problems of the workers.
(C) ensuring workers have access to the most modern technology.
(D) maximizing worker performance.

Leukemia is a blood disease in which white blood cells in the blood or bone marrow reproduce rapidly, interfering with the body's ability to produce red blood cells. Red blood cells are needed to perform vital bodily functions.420 PART 6 Practice ASVAB Exams

10. According to this passage,

(A) white blood cells perform no vital func tion in the body.
(B) no treatment for leukemia exists.

(C) leukemia makes it hard for the body to produce red blood cells.
(D) white blood cells are found only in the Blood.

Questions 11 and 12 are based on the following Passage.

Any discussion of distinctive military capabilities would be incomplete without looking at their relationship to the Joint Service vision of the future. JV 2020 guides all the Services into the next century with its vision of future war fighting. JV 2020 sets forth four overarching operational concepts: dominant maneuver, precision engagement, focused logistics, and full-dimensional protection. Each of these operational concepts reinforces the others. The aggregate of these four concepts, along with their interaction with information superiority and innovation, allows joint forces to dominate the full range of military operations from humanitarian assistance through peace operations to the highest intensity conflict.

11. According to the passage above, which of the following is not an operational concept?

(A) dominant maneuver
(B) focused logistics
(C) high intensity conflict
(D) precision engagement

12. The document discussed in the above passage is primarily about:

(A) military operations of the past.
(B) present military operations.
(C) military operations in the future.
(D) training for future military operations.

Questions 13 through 15 are based on the following Passage.

Genetics is a branch of science dealing with heredity. The field is concerned with how genes operate and the way genes are transmitted to offspring. Subdivisions in the field include cytogenetics, which is the study of the cellular basis of inheritance; microbial genetics, the study of inheritance in microbes; molecular genetics, the study of the biochemical foundation of inheritance; and human genetics, the study of how people inherit traits that are medically and socially important. Genetic counselors are primarily concerned with human genetics. They advise couples and families on the chances of their offspring having specific genetic defects.

13. In the passage above, cytogenetics is defined as:

(A) the study of the psychological impact of genetics.
(B) the study of the cellular foundation of inheritance.
(C) the study of molecular genetics.
(D) the study of human genetics.

14. According to the passage, genetics:

(A) concerns how genes operate and how they're passed along.
(B) is a field of study populated by quacks, fakes, and frauds.
(C) is a field of study only concerned with human genetics.
(D) is a new field of study.

15. According to the passage, it's reasonable to assume that genetic counseling:

(A) is restricted to the very rich.
(B) is used to diagnose diseases.
(C) can be used by parents to learn if their offspring are likely to inherit a disease one of the parents has.
(D) can be used by parents to prevent their offspring from inheriting a specific genetic defect.

Mathematics Knowledge:

1. What is the slope of the line described by the equation y = 4x + 7?
   A. 3
   B. 4
   C. 5
   D. 7

2. If 7x - 2 = 40, what is the value of x?
   A. 5
   B. 6
   C. 7
   D. 8

3. What is the value of x in equation 2x + 5 = 17?
   A. 4
   B. 5
   C. 6
   D. 7

4. A car is sold for $35,000, which is 70% of its original price. What was the original price of the car?
   A. $40,000
   B. $45,000
   C. $50,000
   D. $60,000

5. The length of a rectangle is three times its width. If the perimeter of the rectangle is 40 inches, what is the rectangle's width?
   A. 4 inches
   B. 5 inches
   C. 6 inches
   D. 7 inches

6. What is the sum of the interior angles of a heptagon (7-sided polygon)?
   A. 720 degrees
   B. 900 degrees
   C. 1080 degrees
   D. 1260 degrees

7. If a number is increased by 25%, what is the new number in terms of the original number x?
   A. 1.15x
   B. 1.25x
   C. 1.35x
   D. 1.50x

8. What is the volume of a cube with side length of 4 cm?
   A. 32 cubic cm
   B. 48 cubic cm
   C. 64 cubic cm
   D. 80 cubic cm

9. If the mean of five numbers is 15, what is their sum?
   A. 65
   B. 75
   C. 85
   D. 100

10. What is the slope of the line described by the equation 2y = 5x + 10?
    A. 2.5
    B. 5
    C. 7.5
    D. 10

11. Solve for x: 3x - 2 = 2x + 5.
    A. 5
    B. 6
    C. 7
    D. 8

12. What is the value of 4^3?
    A. 32
    B. 64
    C. 96
    D. 128

13. If a tank can be filled in 4 hours and emptied in 8 hours, how long will it take to fill the tank if both the inlet and outlet are open?
    A. 6 hours
    B. 8 hours
    C. 10 hours
    D. 12 hours

14. What is the square root of 196?
    A. 12
    B. 14
    C. 16
    D. 18

15. Solve for y: 5y - 8 = 3y + 6.
    A. 5
    B. 6
    C. 7
    D. 8

16. A car travels 500 miles on 25 gallons of gas. How many miles per gallon does the car get?
    A. 15 mpg
    B. 20 mpg
    C. 25 mpg
    D. 30 mpg

17. If a square has an area of 144 square units, what is the length of its side?
    A. 10 units

B. 12 units
C. 14 units
D. 16 units

18. What is the result of 15% of 400?
    A. 40
    B. 60
    C. 80
    D. 100

19. What was the original price if a pair of shoes is sold for $120 after a 20% discount?
    A. $100
    B. $140
    C. $150
    D. $160

20. What is the least common multiple of 7, 9, and 11?
    A. 63
    B. 77
    C. 99
    D. 693

21. If the ratio of dogs to cats in a park is 4:3 and there are 21 animals, how many cats are there?
    A. 6
    B. 7
    C. 9
    D. 12

22. What is the sum of the first 6 terms of the arithmetic sequence 3, 6, 9, 12,...?
    A. 36
    B. 45
    C. 54
    D. 63

23. What is the value of x in equation 3x + 4 = 19?
    A. 3
    B. 4
    C. 5
    D. 6

24. If a number is increased by 20%, what is the new number in terms of the original number x?
    A. 0.80x
    B. 1.00x
    C. 1.20x
    D. 1.50x

25. A computer is sold for $1,200, 80% of its original price. What was the original price of the laptop?
    A. $1,000
    B. $1,400
    C. $1,500
    D. $1,600

26. What is the sum of the interior angles of a hexagon?
    A. 540 degrees
    B. 720 degrees
    C. 900 degrees
    D. 1080 degrees

27. If a = 3 and b = 4, what is the value of $a^2 + b^2$?
    A. 9
    B. 16
    C. 25
    D. 34

28. If you roll a fair six-sided die, what is the probability of rolling an odd number?
    A. 1/6
    B. 1/3
    C. 1/2
    D. 2/3

29. Solve for y: 4y + 7 = 3y - 2.
    A. -9
    B. -3
    C. 3
    D. 9

30. What is the value of x in the equation 4x - 3 = 2x + 5?
    A. 2
    B. 3
    C. 4
    D. 5

Electronics Information:

1. What is used to measure the current going through a circuit?
(A) multimeter
(B) amp gauge
(C) current meter
(D) tri-gauge

2. Which of the following isn't a component of a DC motor?
(A) rotor bars
(B) armature
(C) field poles
(D) yoke

3. The television broadcast standard in the United States is:
(A) NTSC.
(B) RGB.
(C) SECAM.
(D) RTSC.

4. In a closed electrical circuit,
(A) one terminal is always positive, and one terminal is always negative.
(B) both terminals can be positive.
(C) both terminals can be negative.

(D) terminals are neither positive nor Negative.

5. Electrical current is counted in what measurement?
(A) hertz
(B) voltage
(C) amps
(D) ohms

6. The following symbol is a/an:

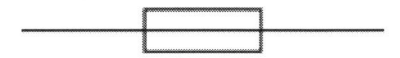

(A) resistor.
(B) fuse.
(C) capacitor.
(D) inductor.

7. what is the specification for an electrical outlet in a bathroom near a sink in the United States?
(A) An outlet must have a childproof cover within 6 feet of a sink.
(B) If within 2 feet of a sink, an outlet must not be GFCI protected.
(C) If within 6 feet of a sink, an outlet must be GFCI protected.
(D) If within 2 feet of a sink, an outlet must also be within reach of the bathtub.

8. The following symbol is a/an:

(A) lamp.
(B) fuse.
(C) inductor.
(D) bell.

9. When a circuit breaker trips, in what position will you find the operating handle?
(A) on position
(B) off position
(C) halfway between on and off
(D) three-fourths of the way between the on position and the off position

10. Which wire is the smallest?
(A) 00 AWG
(B) 4 AWG
(C) 10 AWG
(D) 12 AWG

11. Which of the following is the best conductor of electricity?
(A) plastic
(B) wood
(C) aluminum
(D) copper

12. How many paths of electrical flow can be found in a series circuit?
(A) one
(B) two
(C) two or more
(D) It can't be determined from the information given.

13. A microwave is rated at 1,200 watts. At 120 volts, how much current does it draw?
(A) 1 amp
(B) 10 amps
(C) 100 amps
(D) 1,440 amps

14. Electricians use the term *low potential* to refer to:
(A) electrical circuits with a low potential for overload.
(B) building codes that reduce the risk of fire.
(C) the likelihood of getting a raise this year.
(D) 600 watts or less.

15. Which of the following isn't a conductor of electricity?
(A) water
(B) graphite
(C) gold
(D) glass

16. The ground wire is always
(A) green.
(B) black.
(C) whitish.
(D) blue.

17. What does AM mean?
(A) amp metrics
(B) alien mothers
(C) amplitude modulation
(D) anode matrix

18. Silver is a better conductor than copper. But copper is more often used because of:
(A) the cost of silver.
(B) the brittleness of copper.
(C) the low melting point of silver.
(D) the tendency of silver to tarnish.

19. Electronic circuits that produce high frequencies are called:
(A) amplifiers.
(B) regulators.
(C) transformers.
(D) oscillators.

20. If you plug an appliance designed for AC into a DC power source, the appliance:
(A) will operate normally.
(B) will produce excessive heat.
(C) won't operate.
(D) will explode into tiny pieces.

Auto & Shop Information:

1. If a car uses too much oil, which of the following parts may be worn?
(A) camshaft
(B) connecting rods
(C) fuel pump
(D) piston rings

2. Clean air filters are necessary because:
(A) dirty filters can cause a decrease in fuel mileage.
(B) they remove pollutants, which can decrease engine performance.
(C) they keep the oil from becoming contaminated.
(D) Both A and B.

3. The alternator:
(A) starts the engine.
(B) supplies regulated power to the battery.
(C) connects the ignition system to the engine.
(D) can be used as an alternative to motor Oil.

4. In which automotive system would you find a "wishbone"?
(A) suspension
(B) engine
(C) exhaust
(D) oil pan

5. If the electrolyte solution in a battery is too low, you should add:
(A) sulfuric acid.
(B) antifreeze.
(C) distilled water.
(D) gasoline.

6. What area of your car should be flushed periodically to maintain optimum performance?
(A) exhaust system.
(B) brake system.
(C) cooling system.
(D) ignition system.

7. The primary purpose of a carburetor is to:
(A) maintain engine timing.
(B) regulate oil pressure.
(C) mix fuel and air.
(D) monitor tire pressure.

8. Car restorers often seek NOS parts. What does NOS stand for?
(A) Near Original Specifications
(B) NASCAR Operating Standards
(C) New Old Stock
(D) none of the above

9. To make spark plugs work effectively, the coil and breaker
(A) provide a gap between the electrodes.
(B) ignite the spark.
(C) transfer the electricity to the correct spark plug.
(D) create a very high electrical voltage.

10. Schrader valves can be found in your car's:
(A) tires.
(B) engine.
(C) transmission.
(D) electronic ignition.

11. A bent frame causes:
(A) improper tracking.
(B) auto accidents.
(C) poor visibility.
(D) excessive rust.

12. In the tire designation 205/55 R 15 92 H, what does the "H" signify?
(A) tread type
(B) tire height
(C) maximum sustained speed
(D) turning radius

13. When the tightness of screws and/or bolts is essential, it's best to use:
(A) a screwdriver.
(B) a torque wrench.
(C) tin snips.
(D) a coping saw.

14. Hammer faces are commonly made of each of the following materials EXCEPT:
(A) steel.
(B) brass.
(C) glass.
(D) lead.

15. Hammers, mallets, and sledges are all striking tools, but mallets and sledges don't have:
(A) claws.
(B) metal parts.
(C) as much durability.
(D) heads.

16. Round objects can be measured most exactly using a
(A) rigid steel rule.
(B) folding rule.
(C) set of calipers.
(D) depth gauge.

17. The best chisel to use when making a circular cut in metal is a
(A) masonry chisel.
(B) socket chisel.
(C) butt chisel.
(D) round chisel.

18. A Stillson wrench is a type of:
(A) strap wrench.
(B) hammer.
(C) plumb bob.
(D) pipe wrench.

19. Painting on a surface with too much moisture:
(A) causes no problems.
(B) causes bubbling.
(C) requires an extra coat of paint.
(D) takes longer.

20. A tool used to control the location and/or motion of another tool is called a:
(A) control tool.
(B) jig.
(C) nail.
(D) static rectifier.

Mechanical Comprehension:

1. A simple fixed pulley gives the mechanical advantage of:
(A) 2
(B) 3
(C) 1
(D) unknown

2. The baskets are balanced on the arm in the figure below. If cherries are removed from Basket B, then to rebalance the arm,

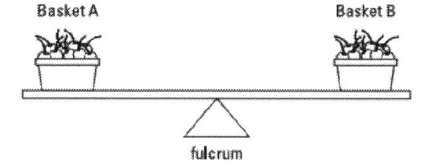

(A) the fulcrum will have to be moved to the right.
(B) Basket B will have to be moved to the right.
(C) Basket A will have to be moved to the left.
(D) Basket A will have to be moved to the Right.

3. If both Wheel A and Wheel B revolve at the same rate in the figure below, Wheel A will cover a linear distance of 12 feet

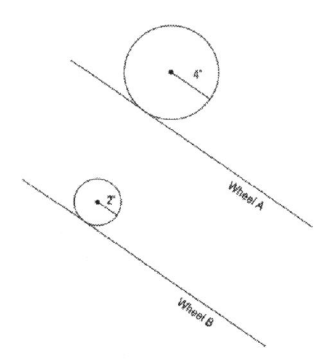

(A) faster than Wheel B.
(B) slower than Wheel B.
(C) in about the same time as Wheel B.
(D) half as quickly as Wheel B

4. If a force of 200 pounds is exerted over an area of 10 square inches, what's the psi?
(A) 10
(B) 15
(C) 20
(D) 200

5. In the following figure, if you move Anvil A toward the middle of the seesaw, Anvil B will:

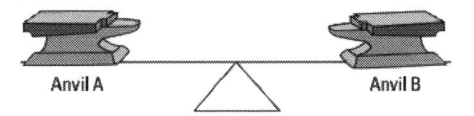

(A) remain stationary.
(B) move toward the ground.
(C) rise in the air.
(D) lose weight.

6. If a ramp measures 6 feet in length and 3 feet in height, an object weighing 200 pounds requires how much effort to move using the ramp?
(A) 200 pounds
(B) 100 pounds
(C) 50 pounds
(D) 300 pounds

7. A micrometer is used to measure
(A) small temperature changes.
(B) changes in psi.
(C) thicknesses to a few thousandths of an inch.
(D) objects invisible to the unaided eye.

8. If the weight is removed from Side B of the seesaw, what happens to the weight on Side A?

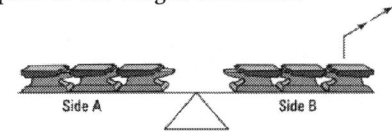

(A) The weight will never move from Side B.

(B) The weight on Side A will move up in the air.
(C) The weight on Side A will move toward the ground.
(D) Nothing will happen.

9. The force produced when two objects rub against each other is called
(A) gravity.
(B) recoil.
(C) magnetism.
(D) friction.

10. Normally, atmospheric pressure is approximately
(A) 14.7 psi
(B) 23.2 psi
(C) 7.0 psi
(D) 10.1 psi

11. For Gear A and Gear B to mesh correctly in the following figure,

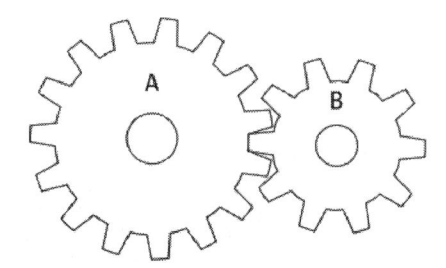

(A) they must be the same size.
(B) they must turn at different rates.
(C) they must both turn in the same direction.
(D) their teeth must be of equal size.

12. Torsion springs:
(A) produce a direct pull.
(B) exert no pull.
(C) produce a twisting action.
(D) coil but do not uncoil.

13. To move a 400-pound crate from the floor of a warehouse to the bed of a truck 4 feet off the ground, the most efficient device to use is a:
(A) lever.
(B) inclined plane.
(C) fixed pulley.
(D) jackscrew.

14. Water in an engine can cause damage in winter weather because:
(A) it can vaporize.
(B) water expands when it freezes.
(C) ice is heavier than water.
(D) cold water creates more steam than warm water.

15. The weight of the load is being carried on the backs of the two anvils shown in the figure. Which anvil is carrying the most weight?

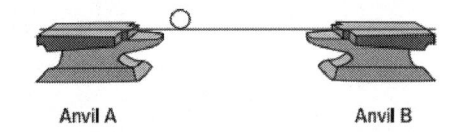

Anvil A   Anvil B

(A) Anvil A
(B) Anvil B
(C) Both are carrying an equal amount of weight.
(D) It can't be determined without more Information.

16. When the block-and-tackle arrangement shown in the figure is used to lift a load, all the following parts remain stationary except:

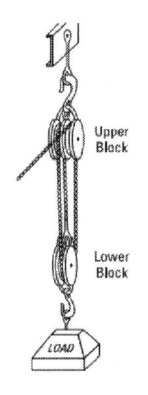

(A) the upper hook.
(B) the upper block.
(C) the lower block.
(D) All the parts move.

17. In the following figure, what effort (E) must be applied to lift the anvil?

(A) 7.0 pounds
(B) 9.0 pounds
(C) 21.0 pounds
(D) 10.5 pounds

18. In the figure below, for each complete revolution the cam makes, how many times will the valve open?

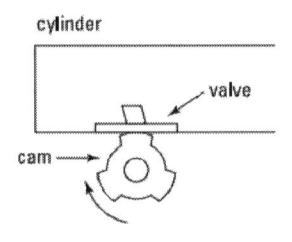

(A) 1

(B) 6
(C) 3
(D) 2

19. In the following figure, assume the valves are all open. Which valves need to be closed for the tank to fill up completely?

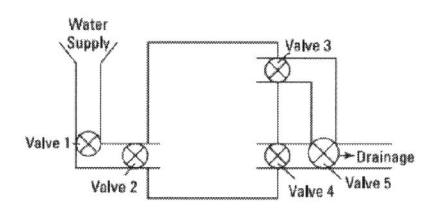

(A) 3 and 4 only
(B) 3, 4, and 5
(C) 2, 3, and 4
(D) 4 only

20. If Gear A turns left in the figure below, Gear B

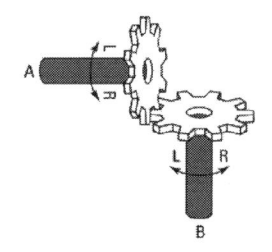

(A) won't turn.
(B) turns left.
(C) turns right.
(D) It can't be determined.

21. If Gear 1 makes 10 complete clockwise revolutions per minute in the figure below, then

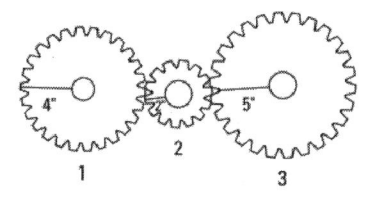

(A) Gear 2 makes 10 complete clockwise revolutions per minute.
(B) Gear 2 makes 20 complete counterclock wise revolutions per minute.
(C) Gear 2 makes 5 complete counterclock wise revolutions per minute.
(D) Gear 3 keeps Gear 2 from making any Revolutions.

22. For the fuel to travel from Reservoir A to Reservoir B, passing through Filters C and D on the way, which valves must be open?

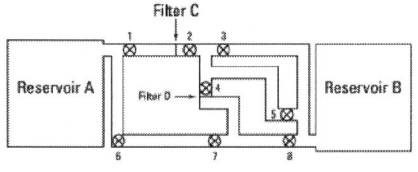

(A) 1, 2, 4, and 8
(B) 1, 2, and 3
(C) 6, 7, and 8
(D) 4, 6, and 7

23. A yellow flame on a gas furnace indicates that
(A) everything is fine.
(B) the fuel-air mixture is too rich.
(C) the fuel-air mixture is too lean.
(D) the gas pressure is too low.

24. If a water tank on a toilet keeps overflowing, the problem is probably a
(A) defective float.
(B) clogged pipe.
(C) crimped chain.
(D) improper seal.

25. In the figure below, the board holds the anvil. The board is placed on two identical scales. Each scale reads

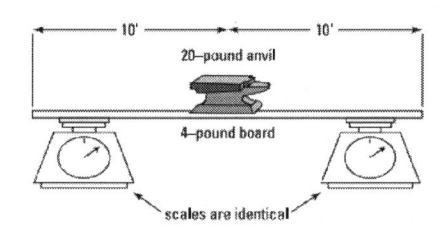

(A) 24
(B) 10
(C) 12
(D) 40

Assembling Objects:

1)

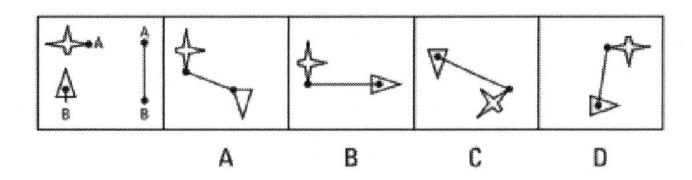

    A   B   C   D

2)

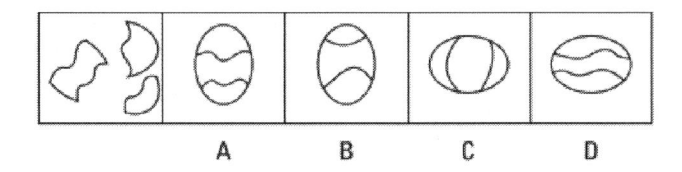

    A   B   C   D

3)

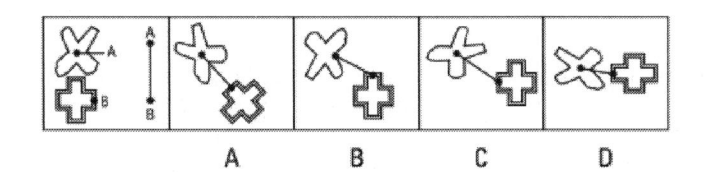

    A   B   C   D

4)

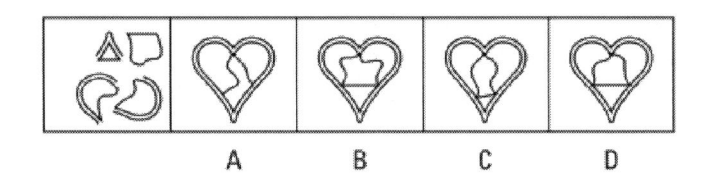

    A   B   C   D

5)

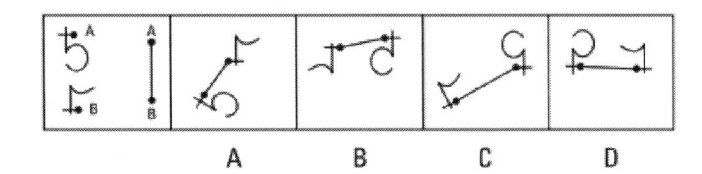

    A   B   C   D

6)

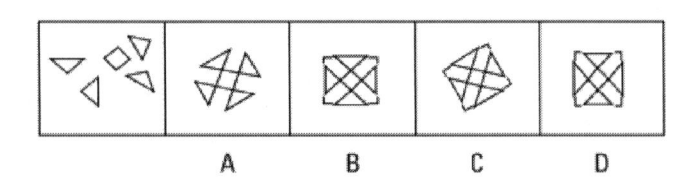

    A   B   C   D

7)

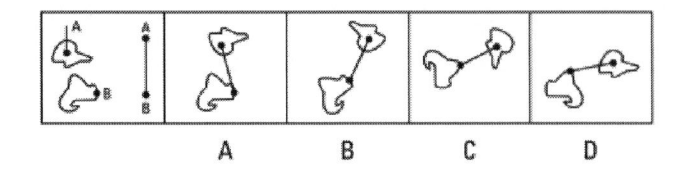

    A   B   C   D

8)

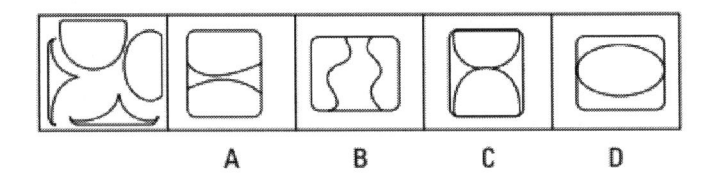

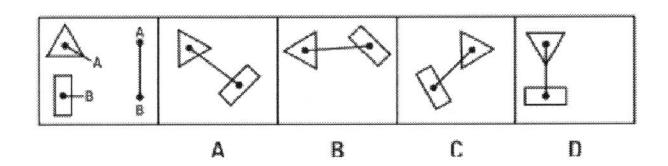

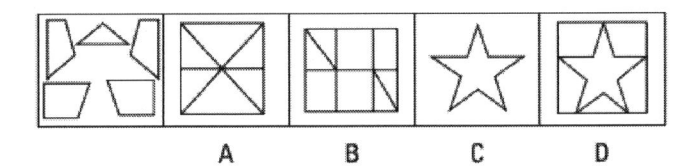

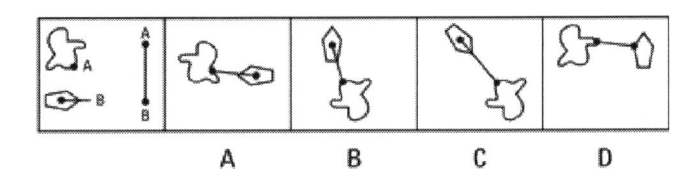

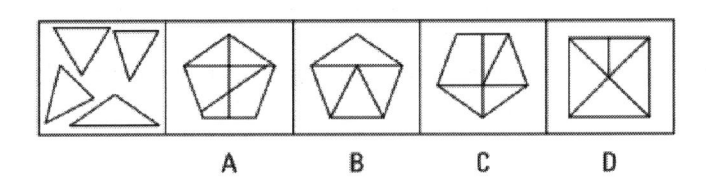

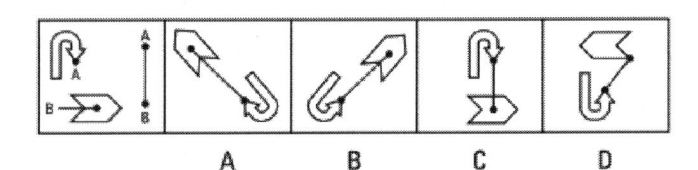

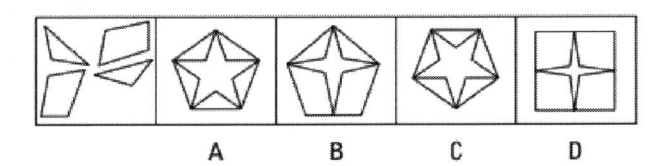

9)

10)

11)

12)

13)

14)

15)

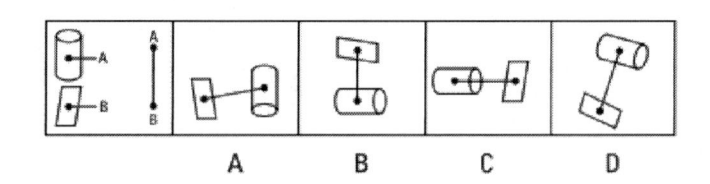

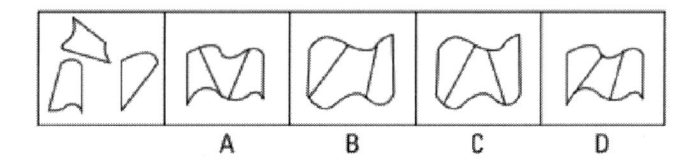

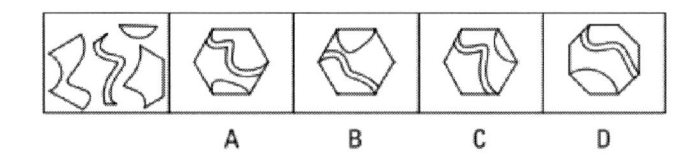

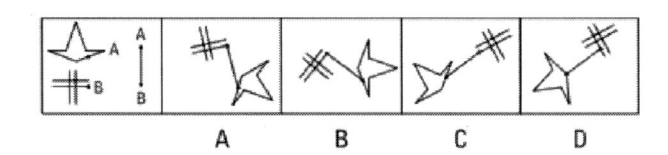

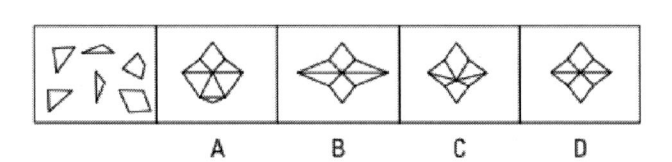

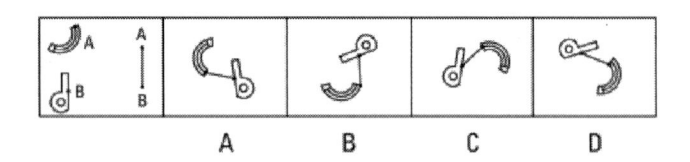

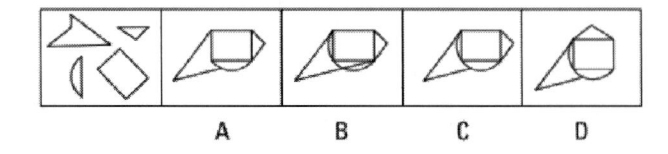

16)

17)

18)

19)

20)

21)

22)

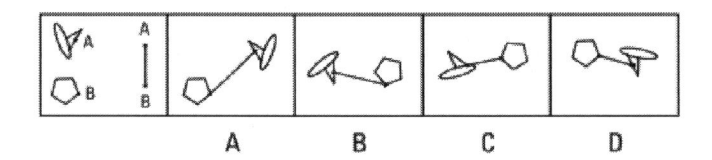

23)

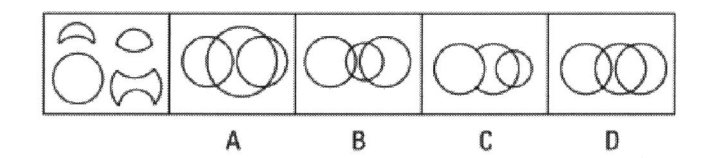

24)

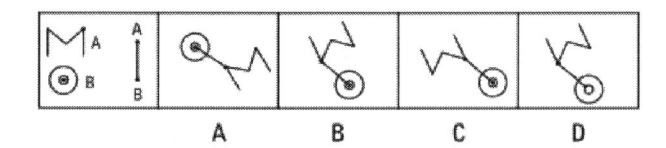

25)

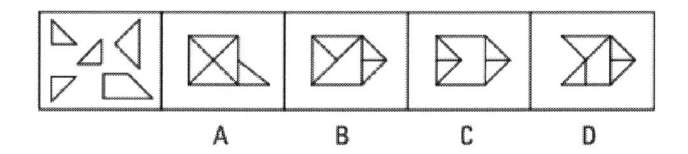

Practice Exam 2: Answers and Explanations
General Science Answers:

1. B.
Metamorphosis is the change an insect's body undergoes.

2. D.
Each increase of 1 in magnitude means a quake is 10 times stronger. Therefore, an earthquake that registers as a 4 on the Richter scale is 100 times stronger than one that rates a 2.

3. C.
Muscles connect to bone with connective tissue called *tendons*.

4. A.
The *stamen* in a flower is the male part.

5. D.
The lungs oxygenate the incoming blood.

6. B.
At more than 4,100 miles, the Nile River is the longest of the rivers listed. Whether it's the longest in the world, as long thought, is now up for debate. In 2007, a group of scientists claimed to have found a new starting point for the Amazon River, which would give the Amazon about a 65-mile lead over the Nile.

7. B.
Physics is the study of matter and its motion, energy, and force.

8. A.
Any meteorologist will tell you that *nimbus* clouds mean rain.

9. D.
Deoxyribonucleic acid—the stuff that genes are made of—is the full term for DNA.

10. B.
Anemos is the Greek word for wind. Therefore, an anemometer measures wind speed.

11. A.
Electrical charges can be positive or negative, so Choice (A) is correct.

12. C.
Jupiter has the most moons in the solar system, with a total of 63 (and possibly more to be categorized in the future).

13. B.
Isaac Newton discovered the law of gravity.

14. D.
The Apollo space program has put 12 men on the moon, with *Apollo* 11 being the first Mission.

15. C.
Omni is an English prefix for "all." Omnivores eat plants and animals.

16. B.
Sharks have only cartilage (in place of bones).

17. A.
The function of the liver is to remove toxins from the blood.

18. C.
Kinetic comes from the Greek word *kinesis,* meaning to move.

19. C.
Terrestrial planets are made up of materials other than just gases.

20. A.
Calcium is a mineral found in dairy, green vegetables, and other foods.

21. D.
The brain of a sperm whale is the largest and heaviest in the Animal kingdom.

22. B.
Stars are classified by makeup, heat produced, and so on. For example, earth's sun is a G type star.

23. C.
As an example, H_2O is two atoms of hydrogen combined with one oxygen atom.

24. C.
A seed is a plant embryo enclosed in a casing.

25. C.
Vernal means spring. *Equinox* means a point in time when day and night are of equal length. This equality occurs twice a year. Therefore, the correct answer is the first day of spring.

Arithmetic Reasoning Answers:

1. **A. 25.**
   Explanation: Let x be the smaller number, and y be the larger number. x + y = 100 and y - x = 50. Solving these equations simultaneously, we get x = 75 and y = 25.

2. **B. 216 miles.**
   Explanation: The car has a fuel efficiency of 120 miles / 5 gallons = 24 miles per gallon. Therefore, it can travel 9 gallons * 24 miles/gallon = 216 miles.

3. **C. 68 mph.**
   Explanation: The train travels 60 mph * 2.5 hours = 150 miles in the first leg, and 80 mph * 1.5 hours = 120 miles in the second leg. The total distance is 150 + 120 = 270 miles. The total time is 2.5 + 1.5 = 4 hours. The average speed is 270 miles / 4 hours = 67.5 mph.

4. **D. 12 inches.**
   Explanation: Let w be the width of the rectangle. Then, the length is 2w. The area of the rectangle is length * width, so 84 = w * 2w, which gives w^2 = 42. Therefore, w = sqrt(42) and the length is 2w = 2 * sqrt(42) = 6 inches. The Length is twice that 2x6=12

5. **B. 400.**
   Explanation: If 30% of a number is 150, the number is 150 / 0.30 = 500. Then, 80% of that number is 500 * 0.80 = 400.

6. **C. $60.**
   Explanation: If the sale price is $45 and this is 75% of the original price, then the original price is $45 / 0.75 = $60.

7. **B. 500.**
   Explanation: The profit per widget is $5 - $3 = $2. Therefore, the manufacturer needs to sell $1000 / $2 = 500 widgets.

8. **B. $15.**
   Explanation: The sales tax is 7.5% of $200, which is $200 * 0.075 = $15.

9. **B. 6:30 PM.**
   Explanation: The clock gains 5 minutes every hour, so it will gain 30 minutes in 6 hours. Therefore, it will show 6:00 PM + 30 minutes = 6:30 PM.

10. **D. $1100.**
    Explanation: The worker earns $20 * 40 = $800 for the first 40 hours. For the next 10 hours, he earns 1.5 * $20 = $30 per hour, totaling $30 * 10 = $300. So, he earns $800 + $300 = $1100 in total.

11. **A. 35.**
    Explanation: There are 0.5% defective widgets produced per day, so there are 0.005 * 1000 = 5 defective widgets each day. In a week, there are 5 * 7 = 35 defective widgets.

12. **A. $1,500.**
    Explanation: The salesman earns a 10% commission on $15,000 sales, which is $15,000 * 0.10 = $1,500.

13. **A. $12,000.**
    Explanation: The car's value after the first year is $20,000 * 0.85 = $17,000. After the second year, it is $17,000 * 0.85 = $14,450. After the third year, it is $14,450 * 0.85 = $12,282.50, which rounds down to $12,000.

14. **C. 45,000 sq ft.**
    Explanation: The area of the football field in square yards is 100 yards * 50 yards = 5000 square yards. To convert to square feet, we multiply by 9 (since there are 9 square feet in a square yard), giving 5000 * 9 = 45,000 square feet.

15. **A. 2.25 gallons.**
    Explanation: The total amount of soda in the box is 24 cans * 12 ounces/can = 288 ounces. To convert to gallons, we divide by 128 (since there are 128 ounces in a gallon), giving 288 / 128 = 2.25 gallons.

16. **B. 7.5 cups.**
    Explanation: The amount of flour required scales with the number of dozens of cookies. If 3 cups of flour are required for 2 dozen cookies, then 5 dozen cookies will require 3 cups * (5/2) = 7.5 cups of flour.

17. **C. 12.**
    Explanation: If the ratio of boys to girls in a class is 3:2, then out of every 5 students, 3 are boys. If there are 20 students in total, the number of boys is (3/5) * 20 = 12.

18. **A. $42,000.**
    Explanation: A 5% raise on a $40,000 salary is $40,000 * 0.05 = $2,000. Therefore, the new salary is $40,000 + $2,000 = $42,000.

19. **B. 20 minutes.**
    Explanation: If the car travels at an average speed of 45 miles per hour, then it travels 45 miles in 60 minutes, so 1 mile takes 60/45 = 1.33 minutes. Therefore, to travel 15 miles it will take 15 * 1.33 = 20 minutes.

20. **B. 100.**

Word Knowledge Answers:

1. C.
Abeyance is a noun that means a state of temporary disuse or suspension.

2. A.
Sturdy is an adjective used to describe something that's strongly and solidly built.

3. A.
Bullock is a noun that's used as a term for a young bull or steer.

4. C.
Brevity is a noun that means the concise, exact use of words, whether in writing or speech. It can also mean shortness of time.

5. B.
Paradigm is a noun that refers to a typical example or pattern of something; it can also mean a model.

6. B.
Facilitated is the past tense of the verb *facilitate,* which means to make something easy or Easier.

7. D.
Quiescence is a noun that refers to a state of inactivity.

8. B.
Spectator is a noun that refers to a person who watches something.

9. B.
Reclined is the past tense of the verb *recline,* which means to lean or lie back in a relaxed position with the back supported.

10. C.
Cited is the past tense of the verb *cite,* which means to quote something as evidence or to justify an argument or statement.

11. C.
Surround is a verb that means to be all around something.

12. B.
Illustrious is an adjective that means well-known, admired for past achievements, and respected.

13. D.
Inhabitant is a noun that refers to a person or animal that lives in or occupies a place.

14. B.
Subterfuge is a noun that means deceit used to achieve a goal.

15. A.
Ghastly is an adjective that refers to something that causes great horror or fear.

16. D.
Rigid is an adjective that means unable to bend or inflexible.

17. A.
Fraternal is an adjective that means of or like a brother (or brothers).

18. C.
Deplore is a verb that means to feel or express strong disapproval of something.

19. B.
Meager is an adjective that means lacking in quantity or quality.

20. A.
Weal is a noun that means well-being, happiness, or prosperity. Don't let Choice (B) throw you off; another definition of *weal* is a red, swollen mark left behind after a blow or Pressure.

21. C.
Guileless is an adjective that means innocent and without deception.

22. C.
Confiscate is a verb that means to take or seize someone's property with authority.

23. D.
Dubious is an adjective that means hesitating, doubting, or not being relied upon.

24. A.
Illusion is a noun that refers to something that is or is likely to be, wrongly interpreted or perceived by the senses.

25. D.
Craving is a noun that means a powerful desire for something.

26. B.
Enmity is a noun that refers to feeling or being opposed or hostile to someone or something.

27. B.
Arbor is a noun that means tree. It can also refer to the leafy, shady area that tree branches or shrubs create.

28. D.
Terminated is the past tense of the verb *terminate,* which means to bring to an end.

29. A.
Buffoon is a noun that means a ridiculous but amusing person.

30. A.
Null is an adjective that means having or associated with the value zero; it also refers to something with no legal or binding force or to something invalid.

31. C.
Indigent is an adjective that means poor or needy. It's also a noun that refers to a needy person.

32. C.
Impertinent is an adjective that means rude or not showing proper respect.

33. D.
Lustrous is an adjective that means shining or glossy.

34. B.
Pardon is a verb that means to forgive or excuse someone or something. It's also a noun that refers to forgiving (or being forgiven) for something.

35. D.
Veracious is an adjective that means speaking or representing the truth.

Paragraph Comprehension Answers:

1. B.
The author is describing a quaint country setting.

2. B.
The passage states that the brooks in the village run south, so the answer is Choice (B). However, the brooks run in an opposite direction (north) a few miles north.

3. B.
The passage states that Panama revolted against Colombia, not that they fought over the canal, so Choice (A) is incorrect. The passage states that foreign companies were unsuccessful in building the canal, so Choice (C) is incorrect. The United States, not Panama, built the canal, so Choice (D) is wrong. In the next to last sentence, the passage states that the U.S. was given rights to the land the canal occupied, making Choice (B) the correct answer.

4. D.
According to the passage, a torn U.S. flag can be professionally mended, but a severely torn flag should be destroyed. The preferred method of destruction is by burning.

5. D.
The passage states that guilds had economic and social purposes, so Choice (A) is incorrect. The passage states that guilds were similar to labor unions, so Choice (B) is incorrect. The passage states that guilds protected merchants and craftspeople; it says nothing about exploiting workers, so Choice (C) is incorrect. The third sentence states that some guilds held considerable economic power, but even small guilds protected members, making Choice (D) the correct answer.

6. D.
According to the passage, it took more than 4 years for the government to believe that anyone had flown a heavier-than-air craft. The historic flight was in December 1903, and the Wright brothers delivered the first aircraft to the government in August 1908, 4.5 years later. The passage supports all the other statements.

7. C.
Freud comments on the characteristics of memory throughout the entire passage.

8. A.
The passage describes how Troy and common weights differ, so Choice (B) is incorrect. Common and avoirdupois are the same system, so Choice (C) is incorrect. A troy ounce is 480 grains, so Choice (D) is incorrect. Choice (A) is the correct answer because the second sentence states that a typical ounce is 437.5 grains, just shy of 438 grains.

9. B.
The passage doesn't address leading by example or using technology by workers, so Choices (A) and (C) are incorrect. Maximizing worker performance results from leadership involvement, not a leadership principle, making Choice (D) incorrect. The correct answer, showing interest in workers' problems, is in the third sentence of the passage.

10. C.
The passage doesn't support Choices (A) or (B). The passage states that white blood cells are found in blood and bone marrow, so Choice (D) is wrong. The correct answer, Choice (C), can be found in the first sentence. The passage states that leukemia interferes with "the body's ability to produce red blood cells."

11. C.
High-intensity conflict is listed as a type of military operation (in the last sentence), not one of the four operational concepts.

12. C.
The JV 2020 guides all the military services with its vision of future warfighting. Although Choice (D) is close, the passage doesn't specifically reference military training.

13. B.
Cytogenetics is the study of the cellular basis of inheritance; the text doesn't support Choices (A), (C), or (D).

14. A.
Nothing in the passage supports Choices (B) or (D). Although human genetics is a vital subfield, nothing in the passage suggests that it's the only concern of geneticists. As the passage mentions, microbial genetics is a subfield in genetics that has nothing to do with humans, so Choice (C) is incorrect. Choice (A) is the correct answer—the second sentence mentions genes and their transmission to offspring.

15. C.
Nothing in the passage supports Choices (A), (B), or (D). Choice (C) is the correct answer because the last sentence in the passage states, "[Genetic counselors] advise couples and families on the chances of their offspring having specific genetic defects." Note it does not state that genetic counselors use genetics to *prevent* offspring from inheriting defects, which is what Choice (D) states, making Choice (D) an incorrect answer.

Mathematics Knowledge:

1. **B. 4.**
Explanation: In the equation $y = 4x + 7$, the coefficient of x (4) is the slope of the line.

2. **B. 6.**
Explanation: To solve for x, add 2 to both sides to get $7x = 42$, then divide both sides by 7 to get $x = 42 / 7 = 6$.

3. **C. 6.**
Explanation: To solve for x, subtract 5 from both sides to get $2x = 17 - 5 = 12$, then divide both sides by 2 to get $x = 12 / 2 = 6$.

4. **C. $50,000.**
Explanation: If $35,000 is 70% of the original price, then the actual price is $35,000 / 0.70 = $50,000.

5. **B. 5 inches.**
Explanation: If the length of the rectangle is three times the width, then the rectangle has dimensions of 3x and x. The perimeter of the rectangle is $2*(3x + x) = 40$, so $8x = 40$ and $x = 5$. Therefore, the width of the rectangle is 5 inches.

6. **B. 900 degrees.**
Explanation: The sum of the interior angles of a polygon with n sides is $180(n - 2)$. For a heptagon, $n = 7$, so the sum of the interior angles is $(7 - 2) \times 180 = 5 \times 180 = 900$ degrees.

7. **B. 1.25x.**
Explanation: If a number is increased by 25% of itself, it is 125% of the original value, which is 1.25x.

8. **C. 64 cubic cm.**
Explanation: The volume of a cube is given by side^3. So, 4^3 = 64 cubic cm.

9. **B. 75.**
Explanation: The mean of a set of numbers is the sum of the numbers divided by the number of numbers. If the mean of five numbers is 15, their sum is $5*15 = 75$.

10. **A. 2.5.**
Explanation: The slope-intercept form of a line is $y = mx + b$, where m is the slope. To get this form from $2y = 5x + 10$, divide both sides by 2 to get $y = 2.5x + 5$. The slope of the line is 2.5.

11. **D. 8.**
Explanation: Subtract 2x from both sides to get x - 2 = 5. Then, add 2 to both sides to get x = 7.

12. **B. 64.**
Explanation: 4^3 is 4*4*4, which equals 64.

13. **B. 8 hours.**
Explanation: When both the inlet and outlet are open, the net rate of filling the tank is the difference between the filling and emptying rates, which is 1/4 - 1/8 = 1/8 tanks per hour. Therefore, it will take 8 hours to fill the tank.

14. **B. 14.**
Explanation: The square root of 196 is 14.

15. **C. 7.**
Explanation: Add 8 to both sides and subtract 3y from both sides to get 2y = 14. Then, divide both sides by 2 to get y = 7.

16. **B. 20 mpg.**
Explanation: The car's fuel efficiency is calculated as the distance traveled divided by the fuel consumed. So, 500 miles / 25 gallons = 20 miles per gallon (mpg).

17. **B. 12 units.**
Explanation: The area of a square is side length squared. If the area is 144 square units, then the side length is the square root of 144, which is 12 units.

18. **B. 60.**
Explanation: 15% of 400 is 0.15 * 400 = 60.

19. **C. $150.**
Explanation: If the shoes are sold for $120 after a 20% discount, then $120 is 80% of the original price. So, the original price is $120 / 0.80 = $150.

20. **C. 99.**
Explanation: The least common multiple (LCM) of 7, 9, and 11 is 99.

21. **C. 9.**
Explanation: If the ratio of dogs to cats is 4:3, then for every 7 animals, there are 3 cats and 4 dogs. Since there are 21 animals, this ratio will occur 3 times. Therefore, there are 3 * 3 = 9 cats.

22. **D. 63.**
Explanation: The first 6 terms of the sequence are 3, 6, 9, 12, 15, and 18. Their sum is 3 + 6 + 9 + 12 + 15 + 18 = 63.

23. **C. 5.**
Explanation: Subtract 4 from both sides of the equation to get 3x = 15. Then divide both sides by 3 to get x = 5.

24. **C. 1.20x.**
Explanation: If a number is increased by 20% of itself, the new number is 1.20 times the original number, or 1.20x.

25. **C. $1,500.**
Explanation: If $1,200 is 80% of the original price, then the original price is $1,200 / 0.80 = $1,500.

26. **B. 720 degrees.**
Explanation: The sum of the interior angles of a polygon with n sides is 180(n - 2). For a hexagon, n = 6, so the sum of the interior angles is 180(6 - 2) = 720 degrees.

27. **C. 25.**
Explanation: Substituting a = 3 and b = 4 into the expression gives 3^2 + 4^2 = 9 + 16 = 25.

28. **C. 1/2.**
Explanation: There are 3 outcomes (1, 3, and 5) that are odd out of 6 total outcomes, so the probability is 3/6 = 1/2.

Practice Exam 2 | 217

29. A. -9.
Explanation: Subtract 3y from both sides to get y + 7 = -2. Then, subtract 7 from both sides to get y = -9.

30. C. 4.
Explanation: Subtract 2x from both sides to get 2x - 3 = 5. Then, add 3 to both sides to get 2x = 8. Finally, divide both sides by 2 to get x = 4.

Electronics Information Answers:

1. A.
A multimeter includes several pieces of test equipment, including an ammeter, which measures inline current.

2. A.
Rotor bars are only on AC induction motors, not DC motors.

3. A.
NTSC stands for National Television System Committee and, although it's gradually being replaced by ATSC (Advanced Television Systems Committee), NTSC is currently the broadcast standard in the U.S. Choice (B) is incorrect because RGB stands for red, green, and blue—the colors of light used to create an image. Although most televisions use this standard, it is not a broadcast standard. Choice (C) is incorrect because SECAM (*Séquentiel couleur avec mémoire,* or sequential color with memory) is a standard used in other countries. Choice (D) is RTSC, which stands for Raytheon Technical Services Company, and is therefore not the correct answer.

4. A.
In a closed circuit, one terminal is always positive, and the other is always negative.

5. C.
Amperes (or amps) are the unit of measure of electric current. Hertz is the unit of measurement of frequency, not current. Current equals voltage divided by resistance. Resistance is measured in ohms. Therefore, neither voltage nor ohms can be the unit of measure for current.

6. B.
The symbol is a fuse. Fuses are designed to *blow* (melt) if the current flowing through them exceeds a specified value.

7. C.
This is a code prescribed by the NEC (National Electric Code). Outlets within 6 feet of a sink need to be GFCI protected for safety reasons.

8. A.
The symbol is a lamp. A *lamp* is a transducer that converts electrical energy to light.

9. C.
Conventional circuit breaker handles have three positions: on, off, and trip. When tripped, the handle moves between the on and off positions. To reset the breaker, move the handle to the off position and then to the on position.

10. D.
The smaller the wire, the larger the number.

11. D.
Plastic does not conduct, and wood is a poor conductor. Aluminum is a good conductor but not better than copper.

12. A.
A series circuit has only one path, so if you break the circuit's path at any point, the electricity stops flowing. An example of a series circuit is a string of Christmas lights that no longer works if a single bulb burns out.

13. B.
I (current) Power (watts) Effort (volts). In this case, $I = 1,200/120 = 10$ amperes.

14. D.

Potential equals voltage; the *low potential* is anything less than 600 watts.

15. D.

Glass is an insulator. Other insulators include plastics, paper, and rubber.

16. A.

Ground wires are always green.

17. C.

Amplitude modulation (AM) was the first type of audio modulation to be used in radio. It works well with high-frequency (HF) and Morse code.

18. A.

Silver is a better conductor, but it's more brittle than copper and more expensive.

19. D.

Oscillators produce high frequencies. An *amplifier* changes the amplitude of a signal. A *regulator* is a circuit that maintains a constant voltage. A *transformer* is a device that changes (transforms) the voltage at its input side to a different voltage on its output side.

20. B.

When DC is applied to an AC appliance, the amount of resistance is less, so more current flows through the wire, and heat build-up.

Auto & Shop Information Answers:

1. D.

Using too much oil is a symptom of piston ring wear.

2. D.

Clean air filters allow your engine to breathe. Air filters remove airborne pollutants from the engine, such as microscopic dirt that can damage cylinder walls, pistons, and piston rings. They also filter out smoke, fumes, and odors. A dirty filter can decrease engine performance and cause a decrease in fuel mileage.

3. B.

The alternator recharges (or supplies power to) the battery.

4. A.

You can find a wishbone in a car's suspension.

5. C.

If the electrolyte solution is too low for a non-sealed battery, you need to add distilled water.

6. C.

For optimum performance, you should flush an automobile's cooling system periodically.

7. C.

In a combustion engine with a carburetor, the purpose of the carburetor is to mix fuel and air.

8. C.

In the world of automobile restorers, NOS stands for New Old Stock.

9. D.

To fire, spark plugs need a very high electrical voltage, supplied by the coil and breaker.

10. A.

The valve you use when filling your car's tires with air is a type of Schrader valve.

11. A.
A bent frame causes improper tracking, meaning the right angle between the centerline and axles is not maintained.

12. C.
The *H* signifies the maximum sustained speed at which the vehicle is stable.

13. B.
Use a torque wrench when specific tightness of screws and/or bolts is required.

14. C.
Hammer faces aren't made of glass.

15. A.
Mallets and sledge hammers don't have claws.

16. C.
A set of calipers is the most accurate widely used tool to measure round objects.

17. D.
You use a round chisel to make circular cuts in metal.

18. D.
A Stillson wrench is a type of pipe wrench.

19. B.
Paint will bubble if the surface you're painting has too much moisture in it.

20. B.
A *jig* is a tool you use to control another tool's location and/or motion.

21. A.
Saws have one fewer tooth than they have points per inch.

22. D.
A mineral aggregate (sand, gravel, or crushed stone), a binder such as natural or synthetic cement, and water make concrete. The mix sometimes contains chemical additives as well.

23. B.
Choice (B) is an image of a hand saw (carpentry saw), which you wouldn't use to cut Metal.

24. D.
The image is of a bolt cutter.

25. B.
The image is of an Allen wrench.

Mechanical Comprehension Answers:

1. C.
A simple fixed pulley gives no mechanical advantage, although it does make work easier by directing the applied force. Therefore, the mechanical advantage is 1.

2. D.
Moving Basket A to the right counterbalances the loss of cherries from Basket B.

3. A.
Wheel B has to make more revolutions to cover the same ground as Wheel A, which covers the distance more slowly.

4. C.
You can calculate psi as Pressure Force/Area. In this problem, P=200/10=20.

5. B.
If you move Anvil A toward the center, Anvil B will move toward the ground.

6. B.
The formula to determine the mechanical advantage of an inclined plane is the Length of the Ramp/Height of the Ramp Weight of Object Effort. Plugging in the numbers gives you

$$\frac{6}{3} = \frac{200}{E}$$
$$6E = 600$$
$$E = 100$$

7. C.
Micrometers measure very small but not microscopic objects.

8. C.
Reducing the weight on Side B will cause Side A to move toward the ground.

9. D.
Objects rubbing together produce friction.

10. A.
Normal atmospheric pressure (the average atmospheric pressure at sea level) is 14.7 psi.

11. D.
Gears of unequal size can mesh properly as long as their teeth are of equal size.

12. C.
Torsion springs coil or uncoil and produce a twisting action, not a direct pull; in other words, torsion springs apply torque.

13. B.
To move a heavy object a few feet in height, the inclined plane is the most efficient device (of those listed) to use. *Note:* The mechanical advantage of an inclined plane is equal to the slope of the plane divided by the height. The longer the slope (compared to the height), the greater the mechanical advantage.

14. B.
Water expands when it freezes, possibly damaging engine components.

15. A.
The load is closer to Anvil A, so it's carrying the greater portion of the weight.

16. C.
All the listed parts remain stationary except the lower block.

17. A.
Apply the leverage formula: Length of Effort Arm Length of Resistance Arm Resistance Force Effort Force:

$$\frac{9}{3} = \frac{21}{E}$$
$$3 = \frac{21}{E}$$
$$3E = 21$$
$$E = 7$$

18. C.
The valve will open each time a high point of the cam hits it. The cam has three high points, so the valve will open three times per revolution.

19. A.
Closing only Valves 3 and 4 keeps the water from leaving the tank.

20. C.
Gears in mesh always turn in opposite directions.

21. B.
If Gear 1 turns at 10 rpm, then Gear 2, which is half the size, turns twice as fast, at a rate of 20 rpm.

22. A.
Opening Valves 1, 2, 4, and 8 allows the fuel to travel through the filters. Opening Valves 1, 2, and 3 doesn't allow the fuel to travel through Filter D. Opening Valves 6, 7, and 8 doesn't allow the fuel to travel through the filters. Opening Valves 4, 6, and 7 doesn't allow fuel to travel to Reservoir B.

23. B.
A yellow flame indicates too much fuel or not enough air. More air should be allowed to enter and mix with the gas. Thus, the fuel-air mixture is too rich.

24. A.
The float measures the water level in the tank. If the tank overflows, the float is probably Defective.

25. C.
The 20-pound anvil and the 4-pound board weigh 24 pounds total or, divided by 2, 12 pounds per scale.

Assembling Objects Answers:

1. B.
Pay attention to where the points are in the original diagram. Point A is located on one of the three spikes in the uppermost figure, and Point B is located near the center of the triangle with a line extending at a 90-degree angle from its shortest side. You can rule out Choices (A), (C), and (D) because of the line coming from Point B.

2. A.
Match each possible answer with the figures in the original diagram. Choices (B), (C), and (D) don't depict the original figures.

3. C.
Point A is attached to a line that leaves the *X*-shaped figure on what's depicted as the right side. Rotate the *X*-shaped figure about 45 degrees and rotate the figure containing Point B 180 degrees, and you'll have Choice (C). Choices (A) and (D) can't be right; Point A's line comes out of the figure at the wrong angle. Choice (B) is incorrect because it shows Point A directly on the shape.

4. D.
Each of the answer choices you're given in this problem is in the shape of a double-lined heart, so look at the shape inside the heart. The only one that matches that extra shape is Choice (D).

5. C.
You can immediately rule out Choices (B) and (D) because the figure containing Point A is inverted in both of them. (Assembling Objects questions feature two-dimensional figures, not three-dimensional figures!) Choice (A) is wrong because the figure containing Point B has the point on the wrong side, which leaves you with Choice (C), a rotated and connected version of the original figures.

6. A.
Choices (B), (C), and (D) are essentially the same, but the lines that enclose each figure are missing from the original image. So that leaves you one choice: Choice (A).

7. B.
Check out the locations of Point A and Point B. Point A comes out of the first amoeba (they look like amoebas, don't they?) through its rounded top. Point B is also attached to the second one in a specific spot. Therefore, the only choice that features accurate depictions of the points is Choice (B).

8. C.
You need to make sure the pieces of this figure fit together correctly, like puzzle pieces. The easiest way may be to count the shapes (four) in the original diagram and compare them to each answer choice. Choices (A), (B), and (D) each only show three puzzle pieces. The only one that shows all four (and happens to be the correct answer) is Choice (C).

9. D.
If you start by paying attention to the points and the lines attached to them, you'll see that Point A's line comes right out of the point of the triangle. You can fast forward through all the other answers and skip to Choice (D), the only one that depicts the line coming from the triangle in the correct place.

10. D.
Do some detective work and compare the shapes in the original diagram with those in the answer choices. Choice (A) shows six shapes, and only one of the shapes depicted matches with one in the original diagram. Choice (B) shows eight shapes, none matching the original shapes. Choice (C) doesn't match any of them, which leaves you with only one possible answer: Choice (D).

11. C.
Look at where the points lie on each of the original figures, and pay attention to where the lines exit the figures, too. You can rule out Choices (A), (B), and (D) immediately because the points are in the wrong places on each figure.

12. B.
Match up the shapes in the original figure and those in each answer choice. Choice (A) depicts a few shapes that don't appear in the original diagram, and so does Choice (C). Choice (D) doesn't add up to what you'd get by rearranging the original figures, but Choice (B) does.

13. A.
Each figure in the answer choices depicts Point A in the correct place, but if you look closely, you'll see that the figure Point A is on is inverted in Choices (B) and (D). You can rule out the final incorrect choice by noting where the line from Point B extends from the shape—it's right through the middle of the angle, which means Choice (C) is wrong. So the only possible answer is Choice (A).

14. B.
Glance at the question, and then look at each answer choice. All the answer choices require the shapes to be identical in size and shape, but that's not the case with the original figures. For example, in Choice (A), you'd need to have five equally sized triangles to make the figure work, and in Choice (C) is the same; it's simply rotated. Choice (D) also requires you to have four identical shapes, leaving Choice (B) as the correct answer.

15. D.
Choices (A), (B), and (C) are wrong, although you'll have to look closely to determine why. Choice (A) is wrong because of the angle at which each line extends from each figure and because the figure that contains Point B is inverted. Choice (B) is incorrect because the figure with Point A in its center is inverted rather than simply rotated, and Choice (C) is wrong because the line coming from Point A comes from the top of the cylinder, not out of its side.

16. D.
This question simply asks you to match the original shapes with those in the answer choices. Compare each, and you'll see that the only figure containing each shape is Choice (D).

17. C.
Put a big mental X over Choice (B), because it's wrong (look at the thinnest wavy shape inside the figure, and you'll see that it doesn't match the one in the diagram). That means Choice (D) is wrong, too. Where did the largest shape in Choice (A) come from? It's not in the original diagram, which means you'd be wrong if you picked that one. But you'd be right if you picked Choice (C) because all the shapes in that figure match those in the original figure.

18. B.
You'll have to wrap your brain around rotating versus inverting for this problem. Choice (D) is wrong because Point A is in the wrong place, so just focus on Choices (A), (B), and (C). Point B is on the right side of the original figure, which looks like an intersection; it's on the southeast side, if that helps you visualize it. Tilt your head if you have to; you'll see that Choice (A) can't be correct because in that figure, Point B is on the northwest side of the intersection. Choice (C) also shows the point in the northwest quadrant, which means only Choice (B) is correct.

19. C.
This puzzle question is tricky, so the best way to solve it is by matching the shapes in each answer choice with those in the original diagram. Start with the kite on the upper right of the original figure (sometimes it's easiest to start with shapes that are clearly out of place among the rest). The only figure it appears in is Choice (C). You could also base your decision on the two thin triangles in the original depiction; they also appear only in Choice (C).

20. C.
Point A on the rainbow (you don't see a rainbow?) lies on the end of the center line in the figure, so forget about Choices (B) and (D). Now look at the lowercase d. In Choice (A), you're looking at a lowercase b, and you know that Assembling Objects questions want you to rotate, not invert, figures; that means the correct answer is Choice (C).

21. A.
The crafty test-question writers who work on the ASVAB will throw plenty of these puzzle problems your way, so use the tried-and-true matching strategy. The only figure among the answer choices that depicts the four-sided, angular shape you see in the original diagram is Choice (A).

22. D.
The pentagon in this figure isn't going to help you; whether you invert it, twist it, or turn it, Point B will always be in the right place. The other figure—the one containing Point A—is what will lead you to the right answer. Choice (C) is a no-go (that's military slang for "failure") because the point is in the wrong place. Choices (A) and (B) are also wrong, because the figure containing Point A is inverted, not rotated.

23. C.
If you look at each answer choice, you'll see an odd shape in each; it resembles an apple core and occurs in the middle of each diagram, where the circles overlap. Compare the similar shape in the original diagram with each answer choice, and you'll see that it appears only in Choice (C).

24. B.
Choices (A), (B), and (D) can't be correct because of where Point A lies on the M-shaped figure, so eliminate those from the pool right away. The only one left is Choice (B), which shows Point A in the correct spot on the figure.

25. D.
The odd figure out among the triangles is the trapezoid, so look through the answer choices to find it. The only one containing it is Choice (D), the correct answer.

Made in the USA
Las Vegas, NV
16 February 2024